The Earth's Crust

Author
Ted Gibb

Program Consultant
Marietta (Mars) Bloch

Australia • Canada • Denmark • Japan • Mexico • New Zealand • Philippines
Puerto Rico • Singapore • South Africa • Spain • United Kingdom • United States

Contents

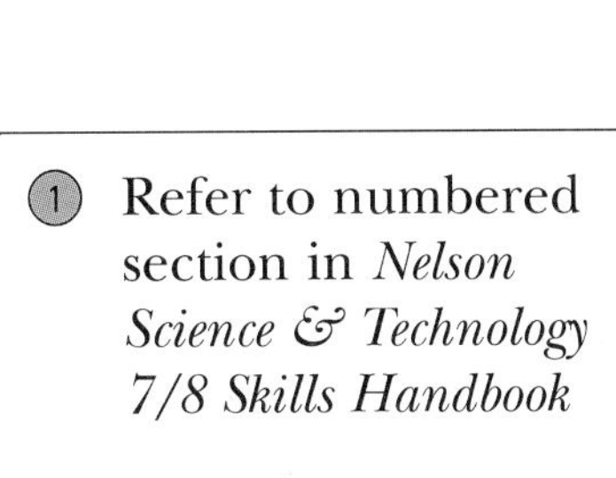

Important safety information

Record observations or data

① Refer to numbered section in *Nelson Science & Technology 7/8 Skills Handbook*

Unit 4 Overview

Rocks hold clues to a mystery that is slowly being unravelled. The mystery is how the Earth came to be as it is, and the rocks in your neighbourhood had a role. Those rocks could be millions of years old, and may have travelled thousands of kilometres from where they were first formed. What events are part of their story? What can we learn from that story to help us survive and live in harmony with our ever-changing planet?

Minerals and Mining

Minerals are the natural substances that form the building blocks of rocks. Human beings extract minerals from the Earth to use as raw materials.

You will be able to:

- understand how minerals formed into rocks over millions of years
- classify minerals according to their characteristics
- describe the many uses of minerals in everyday life
- illustrate how minerals are mined from the Earth
- simulate the economic and environmental issues surrounding the mining of minerals

Weathering and Erosion

Soil consists of weathered rocks and decomposed organic materials. It is essential to the growth of plants and many other living things.

You will be able to:

- describe the different layers of soil
- build models of how soil gets washed away and test ways of reducing that movement
- understand how water moves through the different layers of rock and soil, and the importance of preventing unwanted chemicals from entering the water system
- test soil for qualities that allow a variety of plants to grow in it
- discuss environmental issues related to the reduction of available farmland in Canada

The Dynamic Earth

The Earth's surface is constantly changing as forces uplift, push down, bend, and break the Earth's crust.

You will be able to:

- understand how movements and forces within the Earth's crust gradually created many of Earth's features
- describe how fossils form and what evidence they provide about Earth's history
- identify the three main types of rocks and describe how they formed
- describe how and why volcanoes occur
- build structures that can withstand the vibrations of an earthquake
- predict how earthquakes, mountains, and volcanoes are related to one another
- evaluate evidence that Earth's continents used to be one large continent

Design Challenge

You will be able to ...

demonstrate your learning by completing a Design Challenge.

Model Showing Responsible Use of the Earth's Crust

We depend on mining and farming to provide us with the minerals and food we need to live. But it is essential to carry out these activities in a way that preserves the environment.

In this unit you will be able to design and build:

1 A Responsible Mine
Design and build a model of a working mine that causes minimal environmental damage.

2 A Mine-Tailings Pond
Design and build a model of a pond that safely contains the toxic waste products of a mine.

3 An Erosion-Proof Field
Design and build a model of a field in which the soil resists erosion and water runoff.

To start your Design Challenge, see page 54.

Record your thoughts and design ideas for the Challenge when you see

Design Challenge

Getting Started

Where Do Rocks Come From?

1. All of the metal objects that you see around you originally came from minerals that formed millions, even billions, of years ago in the Earth's crust. Many of these minerals are found deep in the crust, mixed with other minerals. How are they removed from the Earth and purified so that they can be made into things like copper wire?

2. Moving water is very powerful and can carry large amounts of soil and stones with it as it flows downstream. How does moving water affect farmland? Where does soil come from? What's the connection between soil and rock?

3 From the time it formed, the Earth's crust has changed continuously. Volcanoes have erupted, pouring molten lava onto the Earth's surface, which has cooled and formed new rock. Why do volcanoes exist in some places and not in others? What are the forces in the Earth's crust that cause earthquakes?

Reflecting

Think about the questions in 1, 2, 3. What other questions do you have about the Earth's crust? As you progress through this unit, reflect on your answers and revise them based on what you have learned.

Try This: The Great Rock Investigation (6A)

Before beginning this activity, find two or three small, interesting-looking rocks in the schoolyard or on the way home. Each should have some features that make it different from the others. Examine and record the properties of your rocks. In doing so, you might ask the following questions:

1. What colour(s) are they?
2. (a) What do they feel like?
   (b) Are they round or sharp? How do you think they got their shape?
3. Do they look the same throughout, or do they have different types of rock mixed in?
4. Do they feel heavy or light in comparison to their size?
5. Do they have any unusual features (for example, colour, shape, markings)?
6. Do any of your rocks have pieces that sparkle or reflect light?
7. Do your rocks look like most other local rocks? If not, speculate on why they are different, and how they got where you found them.

SKILLS HANDBOOK: (6A) Obtaining Qualitative Data

Earth: A Layered Planet

Just over 4.5 billion years ago, Earth was being formed, as shown in **Figure 1**. According to the most recent astronomers' model, at that time many rocks, large and small, were in orbit around the newly forming sun. As the rocks collided, they joined together to form even larger rocks, and planets like Earth began to form. Because the rocks in space were travelling at high speed, each collision generated huge amounts of heat. This caused the rock to become so hot that it melted. Heavy materials, such as liquid iron, sank to the core of the growing Earth.

As Earth grew larger, more rocks collided with it. Once most of the rocks near Earth were gone, Earth was left about the size it is now. It was also very, very hot.

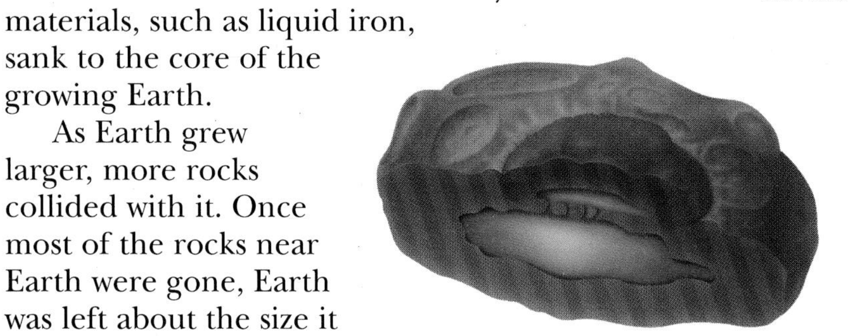

a Two rocks in orbit around the sun collide.

b The collision generates heat, which melts the rock.

c Heavy liquids, such as iron, flow down to the centre of the new body.

Figure 1

Astronomers believe Earth formed from the collision of large and small rocks.

Try This An Eggsact Model?

Is there a good model that can help you visualize the idea of Earth as a layered planet? What about a hard-boiled egg?

- Peel away a piece of the shell of a hard-boiled egg. Compare the thickness of the shell with the rest of the egg.
- Remove the shell from one-half of the egg. Carefully use a knife to cut the egg in half. Use a permanent marker to place a dot at the centre of one-half.

Be careful with sharp objects.

1. Which layer of Earth would the shell represent?
2. Including the centre, how many layers are there in the peeled egg?
3. Does the egg model accurately represent the number of layers of the Earth?
4. Does the thickness of each layer in the egg compare well with the corresponding layer of the Earth?
5. Discuss whether the hard-boiled egg is a good model of the layered Earth.

The Hot Earth

Billions of years later, Earth is still hot. Only the thin crust on the outside has cooled enough to harden into solid rock. Depending on where it is measured, this thin crust varies in thickness from 6 to 64 km—not much compared to the 6400 km from the crust to the core. **Figure 2** shows a model of our layered planet.

1. The Crust

The **crust** is a thin layer of solid rock. The material that makes up the crust tends to be lighter than the materials below—the Earth's crust "floats" on the inner layers.

2. The Mantle

Just below the crust is a hot, partly molten layer called the **mantle**. The mantle is made up of a thick, heavy material. When it cools, it forms rock. The mantle moves sluggishly, like thick syrup.

3. The Outer Core

Toward the centre of the Earth is the core. The **outer core** is a molten mass of mostly iron with some nickel in the mix. Like the mantle, material within the outer core flows.

4. The Inner Core

At the very centre of the Earth is the **inner core**, a large ball of iron and nickel. Despite the heat (almost as hot as the surface of the sun), the inner core is solid, crushed under the enormous weight of the outer core and the mantle.

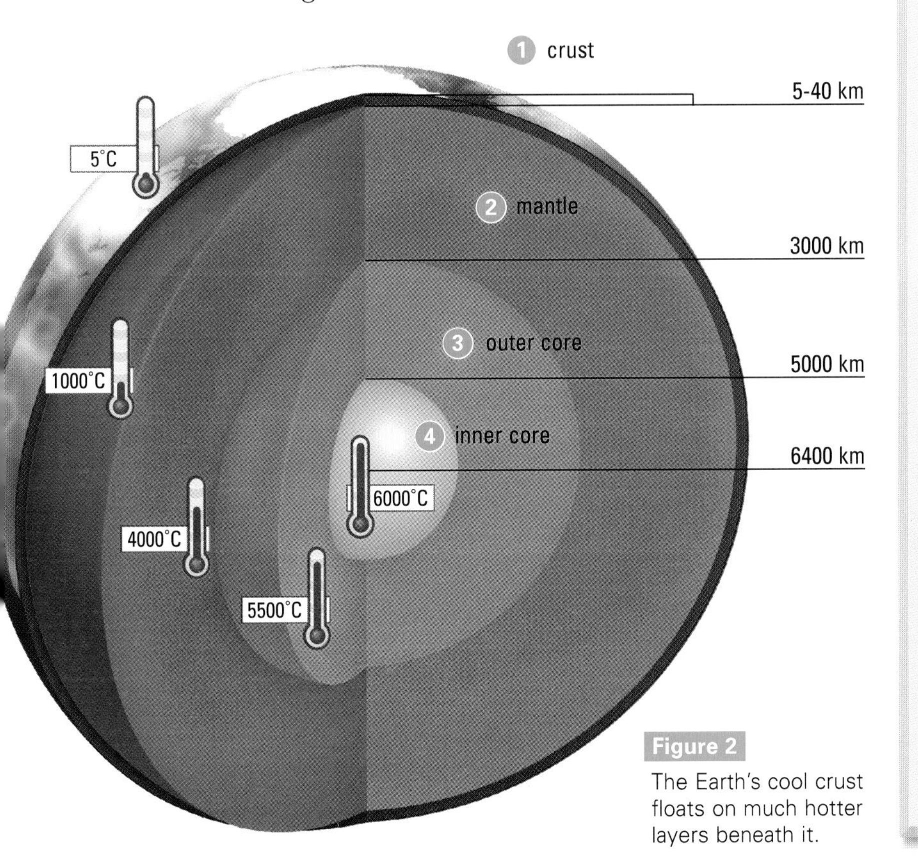

Figure 2
The Earth's cool crust floats on much hotter layers beneath it.

Understanding Concepts

1. (a) Name the layers of the Earth.
   (b) How are they different from one another?
2. Imagine that there was a highway from your school to the centre of the Earth—and that you had a vehicle that wouldn't melt before you got there! At 100 km/h, how long would it take to get from your school to
   (a) the mantle?
   (b) the outer core?
   (c) the inner core?
   (d) the centre of the Earth?
3. Why is the inner core of the Earth solid even though it is so hot?
4. (a) What two metals are found in large quantities at the centre of the Earth?
   (b) Why is there so much more of these metals at the core than in the Earth's crust?

Making Connections

5. Earth is gradually cooling. Draw what you think the Earth's layers might look like in several more billion years.

Reflecting

6. What feature on the Earth's crust gives us an occasional view of how the Earth looks inside?
7. Even on the very coldest days of winter, you can dig a hole through the ice on a lake and always find water. Why doesn't the lake freeze all the way to the bottom?

Minerals: Building Blocks of Rocks

Diamonds, iron, and gold—what are they? They are all minerals, and all come from rock. **Minerals** are the pure, naturally occurring building blocks of rocks. Rocks are made of combinations of minerals. The differences among rocks are due to the minerals they contain.

All minerals are non-living, but a few contain materials that were living things. For example, the mineral calcite, found in chalk, is composed of the shells of tiny organisms that lived in the sea and became fossilized. You will learn more about fossils later.

Because minerals are substances that occur naturally, synthetic substances like steel (iron mixed with carbon) are not minerals.

Minerals and Their Uses

Minerals play a big part in our everyday life (**Table 1**). Gold, iron, and diamonds are all minerals that are used to make a wide variety of objects, including tools, electronic parts, and beautiful jewellery. Gold is used in jewellery because it can be melted and formed into many shapes, and its scarcity gives it value. Like gold, diamonds are also rare. Diamond is prized because it is the hardest mineral known, which makes it useful in tools that cut or scrape, and, of course, it sparkles in light and can be split into smaller pieces, which makes it ideal for jewellery. Iron is rarely used by itself these days but is the main component of steel, used in everything from cars to cutlery.

Minerals Usually Contain Several Substances

Although a few minerals, like gold, may be found in a pure form, most minerals are made of several substances. For instance, iron is found in several different minerals. Two common minerals that contain iron are hematite (iron combined with oxygen) and pyrite (iron combined with sulphur).

Table 1 Some Minerals and the Substances They Contain

Mineral	Location	Valuable substance	Properties of substance	Uses of substance
gold	Northern Ontario	gold	soft, can be shaped, doesn't rust	money, jewellery
diamond	South Africa, Northwest Territories	diamond	extremely hard, reflects light	cutting tools, jewellery
hematite	Northern Ontario	iron	very strong	steel, stainless steel
chalcopyrite	Ontario	copper	conducts electricity, can be easily shaped	wiring, plumbing, coins
uraninite	Ontario	uranium	radioactive	nuclear generation of electricity
chrysotile	Quebec	asbestos	fibrous, doesn't burn	car brake linings, insulation around fireplaces

Minerals Are Crystals

All minerals are pure crystals. Crystals have regular geometric shapes because they are made up of tiny particles connected in a repeating pattern (**Figure 1**). Large crystals form if the mineral cools slowly. Small crystals indicate that the mineral cooled rapidly.

Figure 1
Sulphur crystals. Sulphur is used to make sulphuric acid, which in turn is used to make fertilizer, special steels, plastics, and other chemicals.

Other Ways of Identifying Minerals

- **Lustre** or shininess. Some minerals, like gold, have a metallic sheen. Others, like diamond, look glassy, while asbestos has a dull, fibrous appearance.
- **Cleavage**. Almost all minerals split into smaller pieces with flat surfaces due to their crystal structure. The way they split is called **cleavage**. For example, mica (see **Figure 2**) will always split to form thin sheets. Halite (also called table salt, in **Figure 3**) always splits into cubes. The type of cleavage depends on the shape of the mineral's crystals.
- **Hardness**. Every mineral can make a scratch on other minerals that are softer than itself but cannot scratch a mineral that is harder than itself. Using some standard minerals ranging from very hard to very soft, it is possible to discover quickly the hardness of an unknown mineral. For example talc (see **Figure 4**) is very soft, while quartz (see **Figure 5**) is very hard.
- **Colour**. Although colour isn't as reliable as hardness, it often gives a clue as to a mineral's identity. Gold, for instance, is always yellow. Jade (see **Figure 6**) is usually a shade of green. Quartz is frequently white but can be colourless, violet, grey, or black.

Figure 2
Mica

Figure 3
Halite

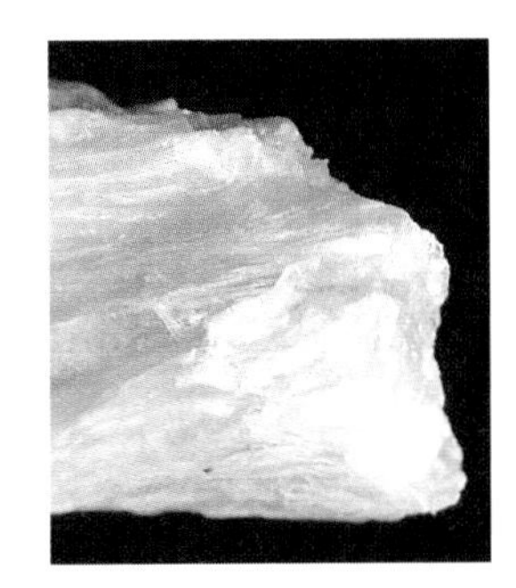

Figure 4
Talc is very soft and has a greasy feel. It is used in cosmetics.

Figure 5
Quartz is hard. It comes in many colours depending on how it was formed. Amethyst, carnelian, agate, and onyx are all forms of quartz.

Figure 6
Jade is used to make jewellery and figurines.

Understanding Concepts

1. Describe the differences between a rock and a mineral.
2. What properties make diamond useful?
3. Describe three properties that all minerals have in common.
4. Why isn't it possible to identify all minerals by colour?

Making Connections

5. Gold, like copper, conducts electricity. Why do you think electrical wiring isn't made of gold?

Exploring

6. People often wear a ring containing their birthstone. Birthstones are minerals that are also called gemstones, or precious or semiprecious stones. A different mineral represents each month of the year. Investigate your birthstone and report on its properties, value, and where it is mined.

Design Challenge

Rocks contain minerals, and minerals may be composed of valuable substances. In your model mine, how could you represent rocks composed of a mineral that contains copper?

How Minerals Are Mined and Processed

Small amounts of many different minerals can be found in most rocks. However, minerals can sometimes be found in high concentrations or **deposits**. For example, the rocks near Sudbury, Ontario, contain huge deposits of minerals that contain nickel and copper. It is these deposits that make mining possible and profitable.

Rock that contains a valuable mineral is called **ore**. High-grade ore contains rich concentrations of a mineral and is the most profitable type of ore to mine. Because it contains less of the valuable mineral, low-grade ore may not be worth mining.

Figure 1

Strip mining is often used when a mineral deposit is discovered near the surface. The top layer of soil and rock, called **overburden**, is removed until the ore is exposed. Then the ore is dug out with large loaders and loaded into trucks.

Searching for Deposits

Many geologists spend much of their lives looking for deposits of useful minerals. Due to the increasing demand for certain minerals, and because known deposits of minerals are being used up, geologists must study the Earth's crust carefully to find new deposits. Once an important deposit is discovered, it is often a complicated process to extract the ore because it may be deep underground, trapped in hard rock. This means that the mine developer must choose the best method for removing the ore from the ground: strip mining (**Figure 1**) or underground mining (**Figure 2**).

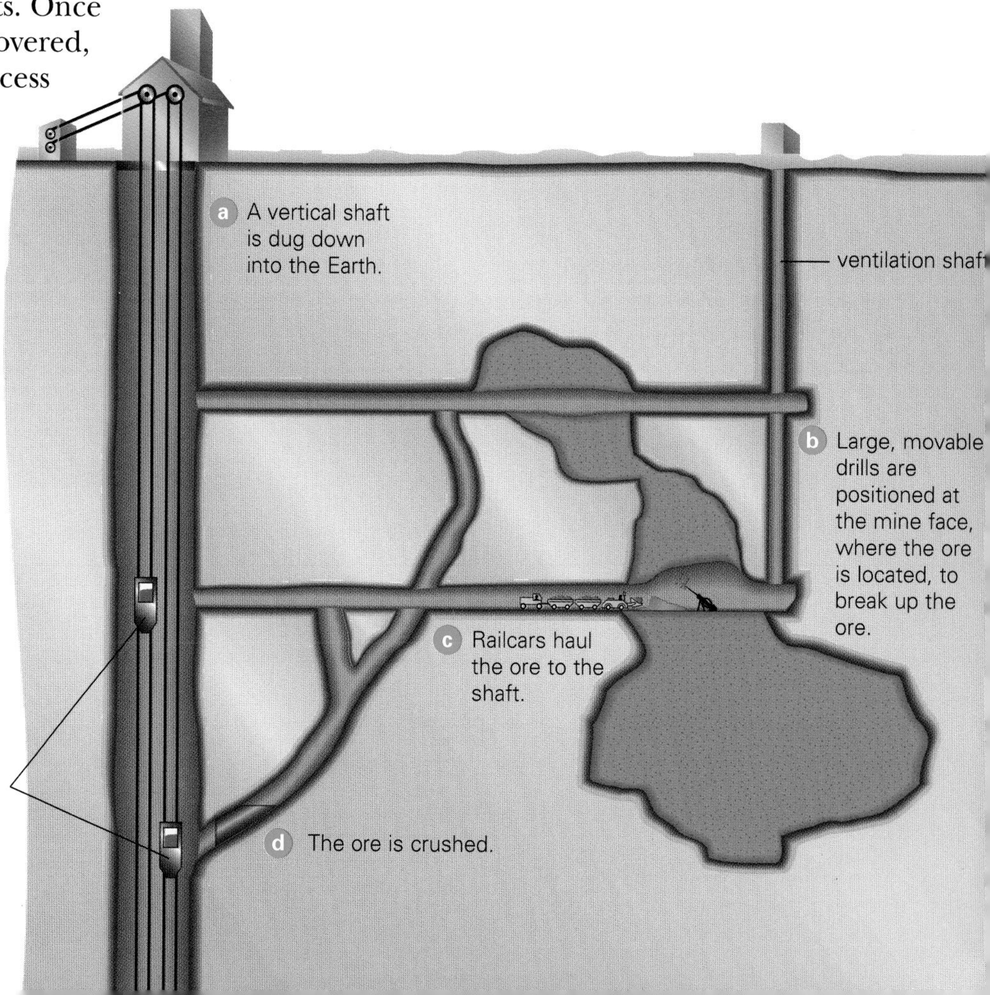

Figure 2

When the mineral deposit is located deep underground, mining directly from the surface is too expensive, so **underground mining** is used.

minerals
sand or chemicals
high temperature smelter
copper

(a) In a flotation tank filled with water and a variety of chemicals, the mineral floats in froth while the unwanted rock sinks.

(b) In a high-temperature furnace, a chemical reaction breaks down the mineral. The reaction requires mixing sand or chemicals with the mineral. This separates valuable substances such as pure copper or nickel.

Figure 3
Separating the valuable mineral from rock

Processing the Ore

All types of ore must be processed to separate unwanted rock from the mineral. The exact process for each type of ore is slightly different, but the key steps are often the same (see **Figure 3**).

These processes use a large amount of water and chemicals, which are recycled as much as possible. However, there are liquid wastes, which are placed in a **tailings pond**. Because these liquids are toxic to people and the environment, the tailings pond must be carefully built so that nothing leaks out (see **Figure 4**).

Figure 4
The tailings pond of this mine is next to the shore, so it is essential that it does not leak.

Deciding When to Close a Mine

For a mine to be profitable, the value of the mineral must be great enough to make it worth the expense of extracting the ore and processing it. Once the high-grade ore runs out, the low-grade ore is usually not worth the cost of processing, and owners will make the decision to close a mine.

Closing a mine also has costs. The mine should be returned as closely as possible to its original natural state, and the liquid waste in the tailings ponds must be removed and disposed of.

Design Challenge

For shallow deposits, miners will create a strip mine. Would a strip mine be suitable for your Design Challenge? Explain. What features would make a tailings pond safe for long-term storage of toxic waste?

Understanding Concepts

1. What is meant by the expressions:
   (a) copper deposit?
   (b) copper ore?
2. Which mining method is used when an ore deposit is far below the surface?
3. Why is the processing of ore an essential part of a mining operation?
4. The mineral deposits for a proposed mine are located as shown in **Figure 5**. Describe how you would extract the ore from each deposit.

Figure 5

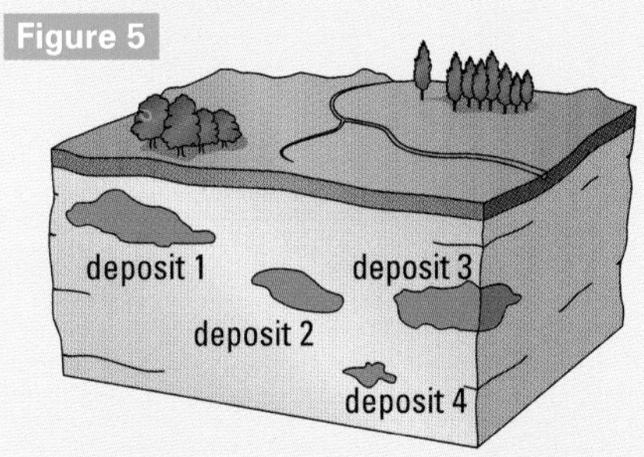

5. A mine operator creates a tailings pond around a natural depression in the land (**Figure 6**). A dam is created to hold in the tailings water by piling overburden at one end of the depression. Is this tailings pond environmentally safe? Give reasons to support your opinion.

Figure 6

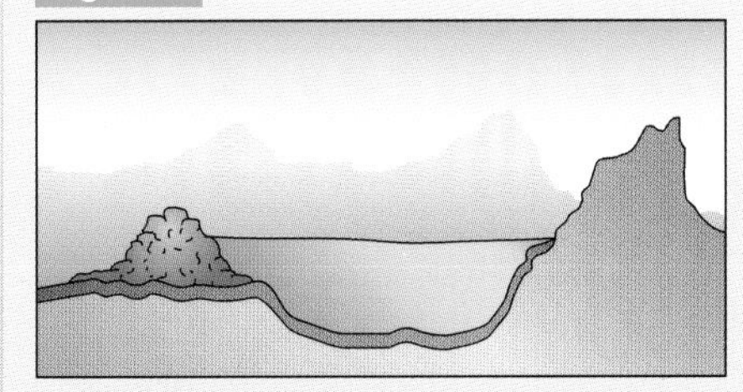

Making Connections

6. How might the varying prices of metals affect a mine operator's decision to close or reopen a mine?

Reflecting

7. Mining is expensive. It requires expensive heavy machinery and many skilled workers. Is there any other source of metals that people could use?

4.4 Inquiry Investigation

SKILLS MENU
- ○ Questioning
- ○ Hypothesizing
- ○ Planning
- ● Conducting
- ● Recording
- ● Analyzing
- ● Communicating

Mining Chocolate Chips

Ore is mined only if it is profitable to do so. After the ore is cut from the rock face, the desired metals or minerals must be separated from the ore (**Figure 1**). There are many ways to do this, but the value of the minerals being mined must always be greater than the cost of the separation process.

In this investigation, you will explore a method to separate chocolate chips from cookies to simulate separation of minerals from ore.

Question

Are some ores more valuable than others?

Hypothesis

Each kind of ore (cookie) will yield a different amount of valuable mineral (chocolate).

Experimental Design

Chocolate chip cookies will be used as a model of ore to explore the costs and benefits of mining.

1 After reading the Procedure, create a method for measuring the amount of chocolate in each cookie.

2 Create a data table to record your observations.

Materials

- 4 chocolate chip cookies (2 each of 2 different brands)
- toothpick
- spoon
- ruler
- several 5-cm lengths of plastic drinking straws

Procedure

3 Use the toothpick to pick the chocolate chips from both the top and bottom surfaces of one of the cookies.

(a) From which surface is it easier to remove the chocolate chips? Why?

4 Use your fingers to break the cookie into smaller pieces so that all the chocolate can be removed with the toothpick.

- Separate the chocolate and the remaining cookie pieces into two piles.

(a) Estimate how much of the cookie was made of chocolate chips. Express your estimate as a fraction.

5 Repeat steps 3 and 4 with a second cookie of the same brand.

(a) How does the amount of chocolate "mined" from the second cookie compare to the amount from the first one?

(b) Using the pieces of drinking straw, make a more accurate measurement of the amount of chocolate in each cookie.

Figure 1

A large part of the expense of mining is removing the valuable mineral from the surrounding rock.

6 Follow steps 3, 4, and 5 using 2 cookies of another brand.

(a) Does one brand of cookie contain more chocolate than the other?

Analysis

7 Analyze your results by answering the following.

(a) Do your results support the hypothesis? Explain.

(b) Why were two cookies of each brand "mined" for chocolate instead of just one?

(c) Was the investigation a fair test of the two brands of cookies? Explain.

Making Connections

1. **(a)** If geologists from a mining company are estimating the richness of an ore deposit, how many samples do you think they should take? Explain.

   (b) Should the samples be taken close together or far apart? Why?

2. **(a)** If the chocolate "mineral" has a market value of $10 per gram, calculate how much the chocolate is worth for each brand of cookie.

   (b) Which brand of cookie "ore" is more valuable?

3. **(a)** Did the chocolate chips separate more easily from one brand of cookie than the other?

   (b) Suppose it costs $20 per hour to pay workers to separate chocolate from the cookie "ore." Explain how the amount of difficulty in removing the chips would affect the value of the ore.

Exploring

4. Invent a machine to separate (3D) chocolate chips from cookies. Draw a diagram of your machine, labelling the parts, and explain how it would work.

Design Challenge

There are many factors that affect how profitable a mine is. You've explored two: the richness (or grade) of the ore and how easy it is to separate the valuable mineral from the ore. How can you represent the ore in your model mine so that it makes sense to mine it?

SKILLS HANDBOOK: (3D) Planning a Prototype

4.5 Explore an Issue

Mining for Minerals

There are many different uses for the minerals that are extracted from the Earth, and the number of those uses is increasing. In Ontario, each person uses 16 tonnes of new material from the Earth every year. As the demand for minerals increases, the Earth's mineral resources are being depleted. Is the solution to open more mines, including inefficient ones with low-grade ore?

In addition to the costs of extracting minerals from low-grade ore, there are also environmental considerations. In the past, mines have damaged nearby land and polluted water systems. Strip mines have an additional problem: if not carefully refilled and landscaped afterward, they can leave an ugly scar on the land on which nothing grows (**Figure 1**).

Restoring the Land

Recently, citizens in mining towns have been demanding that strip-mine operators restore the land when the mine is closed. This involves filling the hole by carefully replacing the rock and topsoil that was scraped off the deposit. The mine operators would also be responsible for replanting native grasses, trees, and shrubs in order to restore the land to its natural state.

Although this kind of restoration is an improvement over simply abandoning a closed strip mine, even with replanting, it does take several years for the natural diversity of plant and animal life to return to the area.

Figure 1
Before beginning a strip mine, extensive environmental planning must be done. (a) Expanding a strip mine. (b) An abandoned strip mine.

Understanding Concepts

1. What kinds of environmental issues need to be considered:
   (a) before a mineral is extracted?
   (b) after a mine closes?

Role Play Mining Economics

Near a town, a deposit of a mineral has been found. The deposit is near the surface, so the company that owns the deposit would prefer to build a strip mine.

The town has a high unemployment rate, and so the citizens welcome development of the mine. However, many of the citizens want a promise from the mining company that the land will be restored after the mine is closed.

The value of the mineral in the deposit is not high at the moment, although it may increase later. The company has said that if it must restore the land completely, this will make the mine unprofitable.

The citizens and their politicians have a difficult choice. They have two options:

1. Wait until the value of the mineral increases so that the mine will become profitable.
2. Offer to subsidize the mine, by paying for restoration of the land, so that the company will be guaranteed a profit.

Sample Opinions of Town Residents

- A subsidy will mean that taxes will increase. We should wait. Jobs will come once it is profitable to build the mine.
- Strip mines disrupt the environment when they are open, and even after a mine site is restored, it takes years for the plant and animal life to return. We should concentrate on finding alternative materials and recycling, and not open any more mines.
- An operating mine will create jobs. We should offer to subsidize the mine and guarantee jobs for local residents. If we wait, the jobs might never come.
- We should subsidize restoration of the land. If we are putting up the money, we can ensure the restoration will be done as we want it done.

What Do You Think? 

Should the town give money to the mining company to cover the extra costs of restoring the land so the mine can begin operating immediately? Or should the town wait for the value of the mineral to rise, with the risk that the mine will never open? Research the arguments further, at the library or on the Internet.

The Roles 

The people listed at right have been appointed to a special committee responsible for making recommendations to the town council about the mine. Choose one of the roles and then prepare your report for the council.

- an executive with the mining company
- an environmental activist
- a local businessperson
- a representative from the provincial environment ministry
- a citizen on a limited, fixed income
- a local politician
- a geologist
- an ecologist

Erosion and Weathering

On your way to school, you may see the same rocks, lawns, and streams every day, and they appear to remain unchanged. But they do change over a long period of time. The changes are caused by erosion. **Erosion** is the wearing away and transport of the Earth's materials. Water, wind, chemicals, and living things all cause erosion.

Figure 1

Water erodes riverbanks.

Erosion by Water

If you live near a lake, you may have noticed that the water appears to be much dirtier after a big storm. This is partly due to large waves hitting the shoreline and knocking soil from the banks into the water. In a river or stream, as shown in **Figure 1**, fast-moving water does the same thing, especially during the spring when melting snow causes the water level to rise and the flow to increase.

On a rocky shoreline, you'll find large boulders, small rocks, pebbles, and sand. Over many thousands of years, the force of the pounding waves breaks rocks into smaller and smaller pieces. Eventually the rocks are broken down into sand.

Erosion by water can occur slowly or suddenly. Over many thousands of years, a mountain stream can gradually wear away even the hardest rocks. But in just hours or days, a swollen river can cause huge clumps of riverbank to fall into the water.

Both the soil washed away by the river and the rocks pounded by waves are undergoing **mechanical weathering**, a form of erosion, shown in **Figure 2**.

Try This: Erosion from Running Water

Make a cone-shaped pile of sand in a shallow pan. Flatten the top, as in **Figure 5**, and place an ice cube carefully on the cone. Allow the ice cube to melt.

Figure 5

1. What changes do you observe in the sand mountain as the ice cube melts?

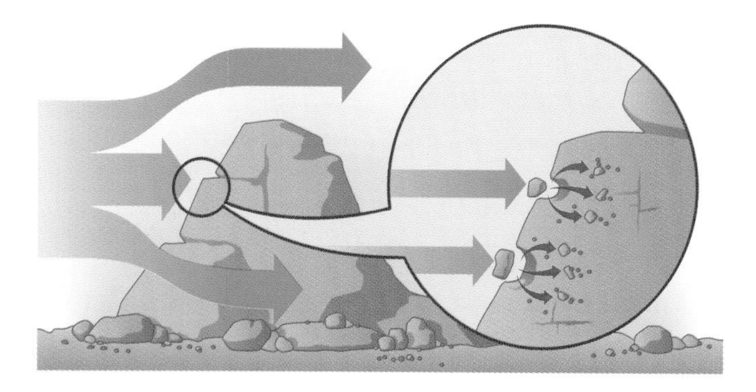

Figure 2

Mechanical weathering by water or wind. As sand particles carried by water or air hit a rock, they wear away the rock.

Erosion by Wind

Wind is another force that can cause mechanical weathering. The devastating dust storms in the Prairies during the severe droughts in the 1930s demonstrated the wind's power. Due to a lack of rain, the rich surface soil became very dry. This allowed the strong winds sweeping over the surface to pick up the light, dry soil particles and blow them many kilometres away. In many places, the rich soil was completely blown away, leaving behind layers that were unsuitable for growing crops. Many farmers were forced to abandon their farms.

Wind not only erodes soil, but it can also erode rock. Over time, windblown particles of soil or sand can wear away rock, much as sandpaper wears away wood. Some layers of rock are softer than others and wear away faster, producing interesting rock formations like the Hoodoos in Southern Alberta (**Figure 3**).

Figure 3

The Hoodoos in Alberta were formed by wind erosion. Wind has worn away the rock by driving soil and sand particles against it for thousands of years.

Erosion by Ice: Nature's Bulldozer

One of the more dramatic causes of mechanical weathering is glaciers. Although they seem to stay in one place, glaciers actually move very slowly downhill due to their immense weight.

As these huge masses of ice move down a mountainside, they slowly scrape the rock layer below, causing it to break up, as shown in **Figure 4**. The erosive force of a moving glacier is so great that it can even scrape a U-shaped valley from the rock beneath it.

Once a glacier starts to melt, it leaves the broken rocks it eroded in enormous piles, called **moraines**. Moraines are often as high as 100 m. Long, narrow moraines may be deposited along either side of a melting glacier or at the front of a glacier, where it is melting and receding. Moraines can be found in many places in Ontario, but they're especially noticeable on the northern shores of Lake Ontario and Lake Erie. Those hills, and the gravel they are mined for, were left by the last great glacier.

(a) One of the glaciers of the Columbia Icefield in the Rockies.

(b) During the last Ice Age, a huge glacier covered most of the continent. You may find a rock like this, with marks that have been cut into it by moving ice.

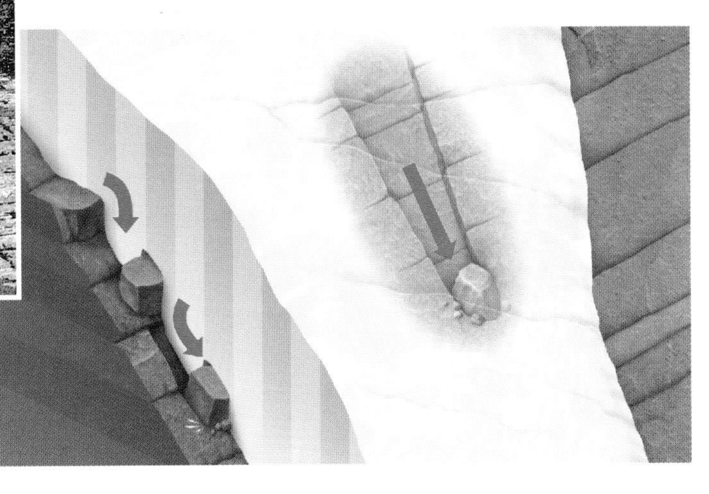

(c) Rocks are broken off as the glacier moves.

Figure 4

Slowly moving glaciers cause massive erosion of the Earth's surface as their immense weight scrapes along the ground.

Erosion by Ice: Nature's Chisel

Another type of mechanical weathering occurs when rainwater seeps into cracks and pores in rock surfaces. When the temperature dips below the freezing point, the water changes to ice. Water expands as it freezes. As the water expands, it puts pressure on the walls of the crack, forcing it to widen. Eventually the rock may split down the crack, or pieces may break off, as shown in **Figure 6**.

Figure 6

Water expands when it freezes, putting pressure on surrounding rock.

Erosion by Living Things

Lichen, a form of living thing that is part fungus and part plant, grows on rocks. Lichen uses the minerals in the rocks as a source of nutrients. It produces an acid that dissolves the rock on which it's growing. The acid wears the rock down. When the lichen dies, it leaves a thin layer of material—soil—in which other plants can grow.

Cracks formed by expanding ice can also be a home for plants. The wind may deposit soil particles in a crack, or soil may be formed there by lichen. Then, if windblown seeds also get carried into the crack, the small amount of soil may support the growth of plants. The roots of these plants will often penetrate deep into the cracks in their search for water and nutrients, causing the cracks to deepen and widen, as shown in **Figure 7**. Again the rock may split or pieces may break off.

Weathering due to the action of living things is called **biological weathering**.

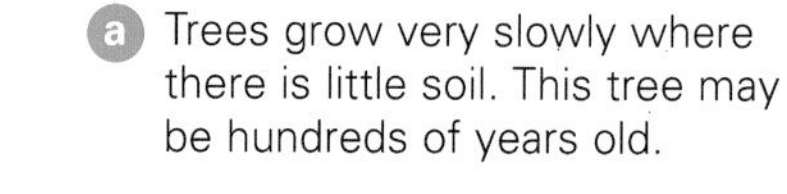

a Trees grow very slowly where there is little soil. This tree may be hundreds of years old.

b As the tree's roots grow, they split the rock.

Figure 7

Trees and other plants cause biological weathering with their roots.

Try This: Ice versus Water

Use a permanent felt-tip marker to mark a horizontal line near the top of 4 to 6 compartments in an ice-cube tray. As shown in **Figure 8**, carefully add water to each compartment up to the line you drew. Place the tray in the freezer and leave it until solid ice cubes have formed. Remove the tray and examine the cubes.

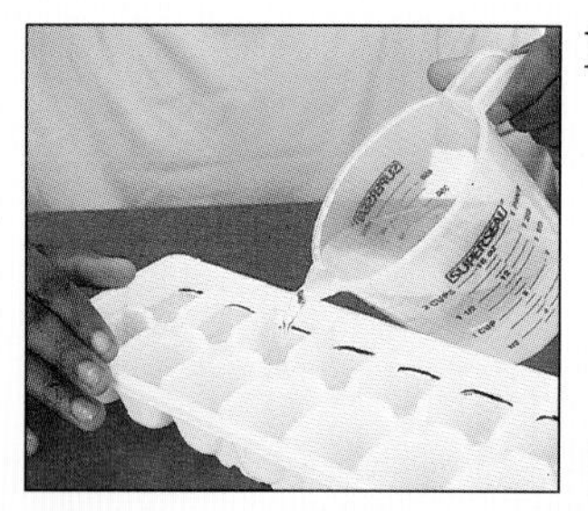

Figure 8

1. How does the size of the solid cube compare to the amount of liquid that was in each compartment?

Erosion by Chemicals

Rocks can also be broken down by chemical weathering. **Chemical weathering** occurs when water, air, and other materials react with the rocks, changing the substances that make up the rocks.

Water causes mechanical weathering, and it is also the main cause of chemical weathering. As shown in **Figure 9**, when water passes over or through rock, it dissolves certain minerals and carries them away, changing the makeup of the rocks.

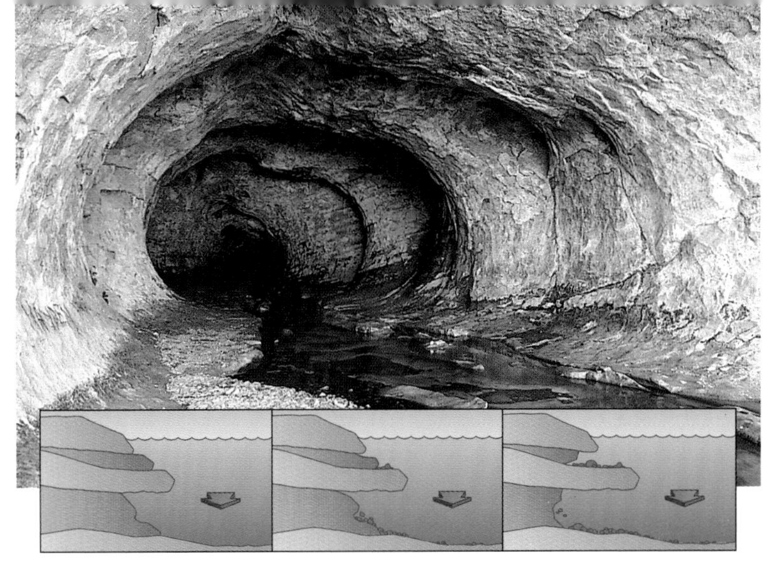

Figure 9

Water dissolves the mineral calcite in limestone, sometimes forming large underground tunnels.

Another example of chemical weathering is acid precipitation (either snow or rain) caused by pollutants and natural acids in the air. Acid rain dissolves more minerals than ordinary water. An example is the limestone rock used for statues and other monuments and buildings (**Figure 10**). Over many years, the mineral calcite in this rock dissolves as acid rain pours over it, causing the rock to crumble.

Figure 10

The chemical weathering of ancient statues has been speeded up by the modern pollutants in acid rain.

Chemical weathering and biological weathering are sometimes difficult to tell apart. Is the acid made by lichen an example of chemical weathering or biological weathering?

Weathering: Fast or Slow

Weathering can occur quickly or slowly depending on several factors. A small rock will often weather faster than a large rock because smaller rocks have proportionately more surface exposed to the forces of erosion.

The minerals in a rock will also affect the rate of weathering. Granite, for example, is much more resistant to chemical weathering than limestone or marble because the minerals in granite do not dissolve easily in water.

Climate is also important. For example, during our cold winters, mechanical weathering by ice occurs. On the other hand, areas near the equator, where there is no ice, are more likely to suffer from huge storms, which can do more mechanical weathering in a few hours than ice in the North could do in a century.

Understanding Concepts

1. Develop a concept map about (9E) erosion.
2. Make a chart explaining the differences between mechanical, biological, and chemical weathering.
3. Water is an agent of both mechanical and chemical weathering. Explain.

Making Connections

4. Some scientists list gravity as one of the causes of erosion because sloping land erodes more quickly than flat land. Explain why.
5. Name a natural or human-made (6C) feature near where you live that shows the effects of erosion. Make a diagram to show the kinds of weathering that are occurring. Has an attempt been made to prevent the weathering? If so, describe it.

Exploring

6. Use the Internet to take a (4A) virtual field trip to a glacier in the Rockies. Is the glacier advancing or receding? Use web site evidence to support your answer.

Design Challenge

What kinds of erosion would you expect to act on a farmer's field? List the possibilities.

SKILLS HANDBOOK: (9E) Graphic Organizers (6C) Scientific & Technical Drawing (4A) Research Skills

Learning About Soil

Over time, rocks can be broken down into smaller pieces by weathering. These particles of rock are the beginning of a process that leads to the formation of soil in which plants can grow. As soil develops, it creates a series of layers called **horizons**, as you can see in **Figure 1**. In each deeper layer, the size of the rock particles tends to increase.

1. Litter

The surface of soil is usually covered with leaves, broken branches, and fallen trees. This layer, known as the **litter**, keeps the ground damp by preventing too much water from evaporating.

2. Topsoil

Beneath the litter is a layer of **topsoil**. Topsoil usually contains dark, decaying plant and animal matter called **humus**. Humus is important because it contains the rich supply of nutrients and minerals that new plants need for growth.

3. Subsoil

The **subsoil** contains larger pieces of rock and clay. It is usually a lighter colour because it contains little humus.

4. Bedrock

A layer of solid, unbroken rock called **bedrock** marks the dividing line between soil and rock. The bedrock may be under soil, but it is still subject to biological weathering from plant roots.

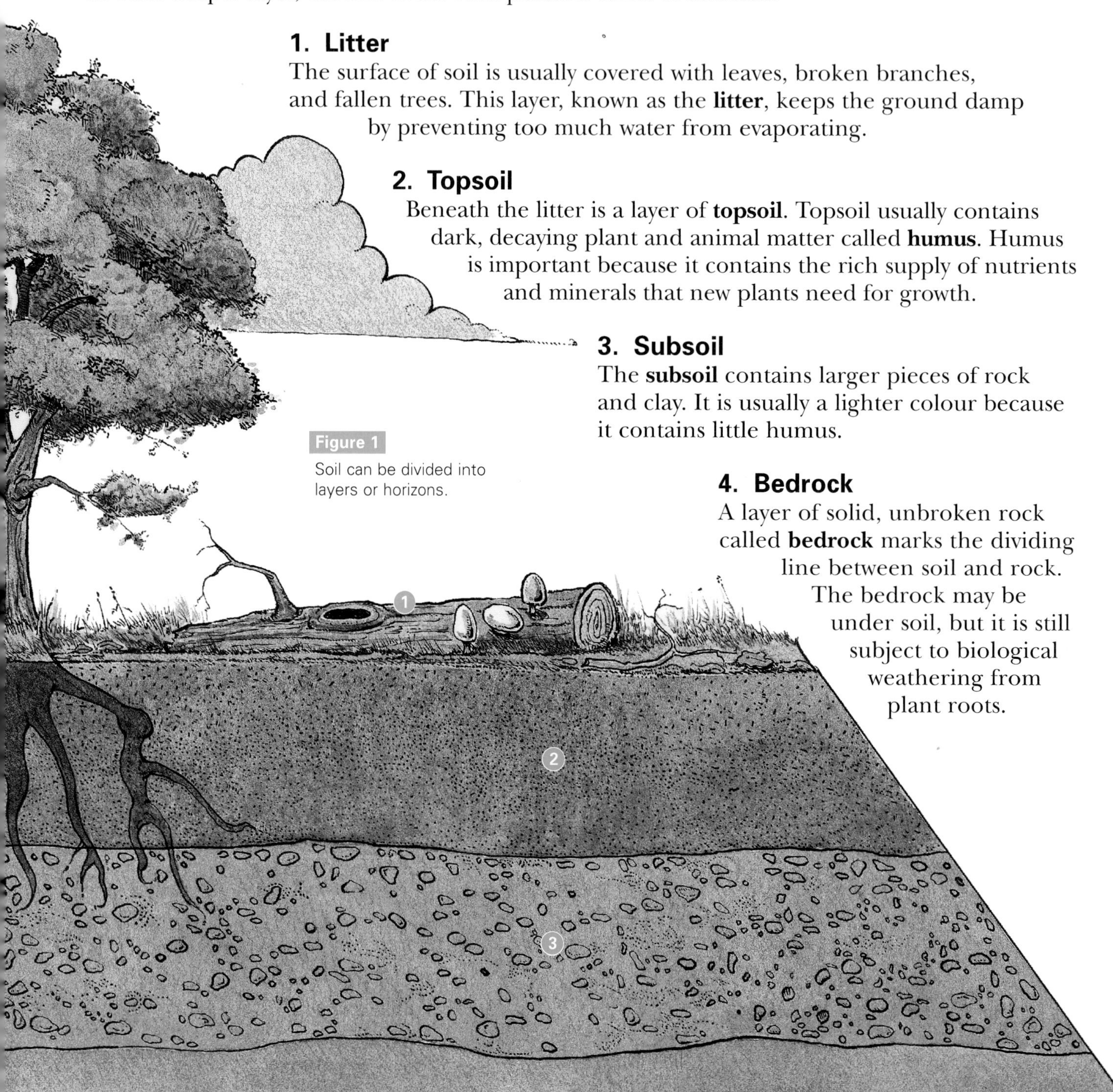

Figure 1
Soil can be divided into layers or horizons.

Is Soil the Same Thickness Everywhere?

As bedrock is weathered, small rocks break off, deepening the subsoil. The subsoil is also being weathered by plant roots and small burrowing animals, such as moles and worms, which bring humus down into the subsoil. As a result, the top of the subsoil slowly becomes topsoil. This process takes time—thousands of years.

Depending on how long the soil has had to form, how much material was left by the glaciers, and the amount of erosion that has occurred, the soil has different thicknesses in different parts of the country. In southern Ontario, where great trees and countless animals have lived, died, and contributed their bodies to soil humus for 10 000 years, soil layers tend to be deeper. In northern Ontario, where the glacier of the last Ice Age lasted much later, there has been less time for soil to form. It is also cooler in the north, so plants grow more slowly and there is less biological weathering. Near the glaciers of the Columbia Icefield in the Canadian Rockies, little pine trees can be seen growing in rocky soil that is just a few centimetres deep. There has been little time there for soil formation.

Design Challenge

In addition to forming humus, trees and other plants have another important function in soil: their roots help soil resist erosion by water and wind. How could you use trees in your model field?

Understanding Concepts

1. Why is soil important?
2. What features distinguish each soil layer from the others?
3. Describe how soil is formed.

Making Connections

4. (6C) Examine the picture of the northern forest shown in the Try This. Draw a diagram showing what you would expect to happen to soil formation if the trees were cut down and removed.

Exploring

5. (4A) If there is little or no oxygen in the soil, decayed plant matter will form a material called peat instead of humus. Research some of the important uses of peat.

Reflecting

6. The treeline is the northern boundary at which trees are able to grow. Using a map showing the treeline across Canada, explain why this boundary might move when considering changes in climate and soil layers.

Try This: Soil Horizons

Compare the soil horizons in **Figure 2**. Which soils do you think would be best for growing crops? Give your reasons.

Figure 2

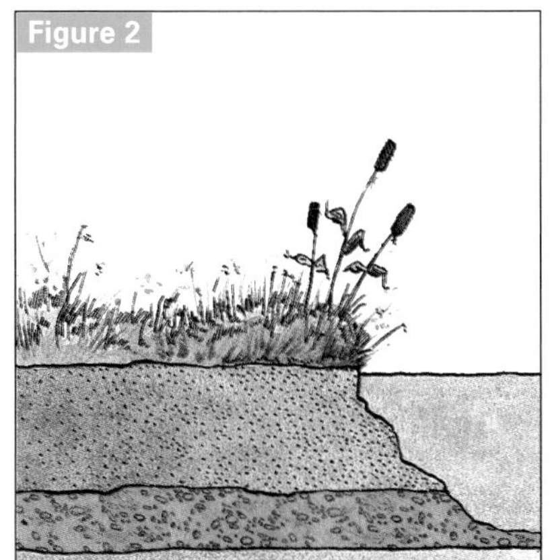

grassland

southern forest

northern forest

4.8 Inquiry Investigation

SKILLS MENU
- ○ Questioning
- ● Hypothesizing
- ○ Planning
- ● Conducting
- ● Recording
- ● Analyzing
- ● Communicating

Components of Soil

You've probably noticed that soil seems to be different when you dig in different areas. Some soils are soft, dark, and rich; others are thick and grey. In some soil, many different kinds of plants grow easily; in others, only very hardy plants can grow. Topsoil also contains particles of various sizes. In this investigation, you will explore how these particles affect soil quality.

Materials

- safety goggles
- apron
- trowel
- plastic bag
- 100-mL graduated cylinder
- clear plastic jar or bottle with lid
- 50-mL beaker
- plastic tablespoon

Do not handle the soil. Soil may contain sharp objects, such as glass or nails.

Question

What factors make topsoils differ?

Hypothesis

1 (2C) Based on what you know about soil, write a hypothesis that explains why topsoils differ.

Experimental Design

In this investigation, each group will examine a sample of topsoil. You will compare your results to those of the other groups and analyze differences.

2 (6D) After reading the Procedure, design a chart to record all of your observations.

Procedure

Part 1: Viewing Particles in Topsoil

3 Find an area where you can remove some soil.

(a) Describe the soil. Is it easy to dig? What colour is it? Are there many small rocks in it or only a few? What kinds of plants grow in the soil?

4 Using a trowel, place 1 scoop of soil into a plastic bag.

(a) Record where the sample was taken.

5 Fill a clear plastic jar one-quarter full with soil from your sample.

- Add water until the jar is three-quarters full.
- Screw the lid on and shake the jar well.
- Leave the jar for 2 min or until the contents settle.

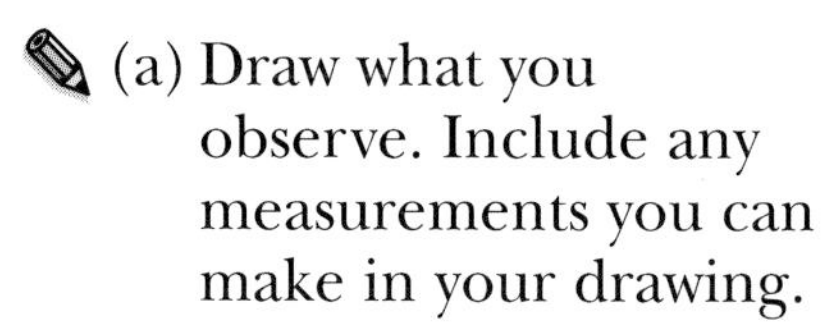

(a) Draw what you observe. Include any measurements you can make in your drawing.

(b) Are the larger particles found near the top or bottom of the jar?

SKILLS HANDBOOK: (2C) Predicting and Hypothesizing (6D) Creating Data Tables

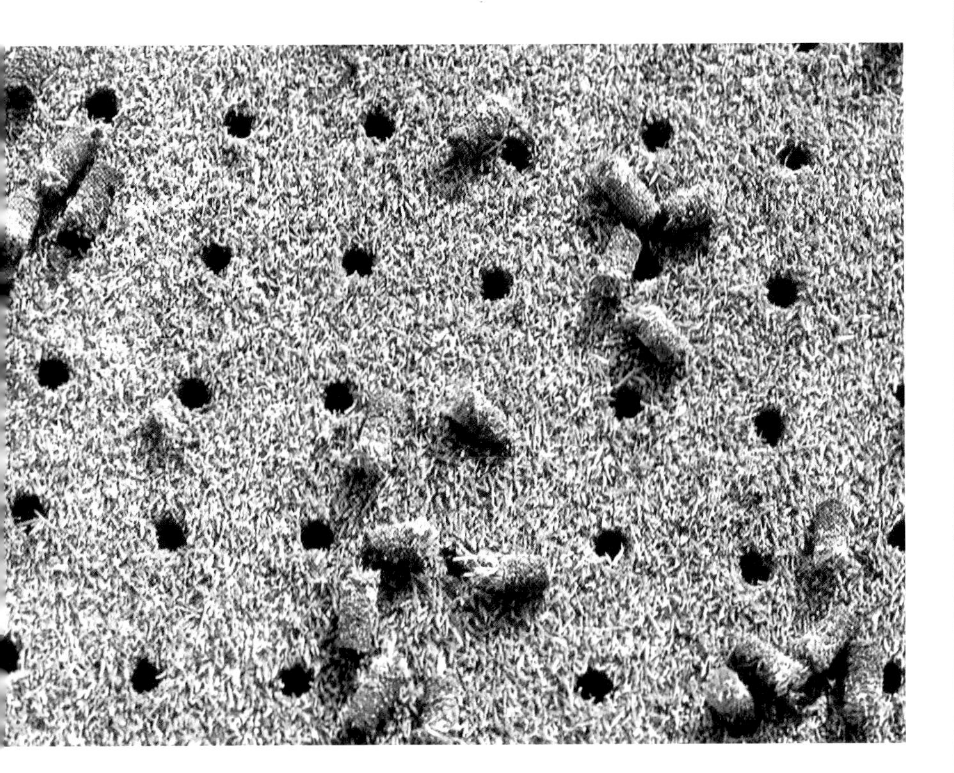

Figure 1

Gardeners and groundskeepers sometimes go over their lawns with a machine called an aerator. Can you explain why?

Part 2: Measuring the Air Content

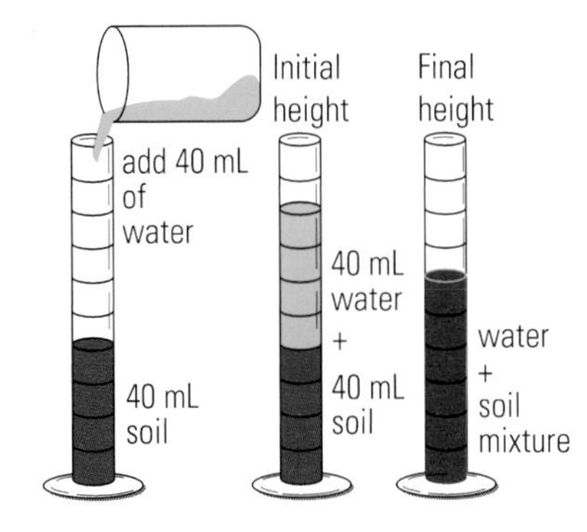

6 Put on your safety goggles.

- Using the graduated cylinder, measure 40 mL of water and pour it into a small beaker.
- Dry the graduated cylinder and, using a spoon, measure 40 mL of your soil sample in the graduated cylinder.
- Slowly pour the 40 mL of water onto the soil in the graduated cylinder.

(a) Record the final level of the soil and water.

(b) Did you observe any bubbles coming from the soil when the water was added? What does this tell you?

(c) Calculate the change in volume of the water and soil when they were combined:

Initial volume = 80 mL
Final volume =
Change in volume =

(d) Why is the final volume less than the initial volume?

Analysis

7 Analyze your results by answering these questions.

(a) In step 5, did all the groups in your class get the same results? Describe any differences.

(b) In step 6, which of the soil samples in your class held the most air? Which held the least air?

(c) Compare all the groups' descriptions in step 3 and the results from steps 5 and 6. What do you notice?

(d) Do the results of this investigation support your hypothesis? Explain.

Making Connections

1. **(a)** Why might plants grow best in soils that have air spaces?
   (b) Use your answer to explain why the plugs of earth have been removed from the grass in **Figure 1**.

Exploring

2. In this investigation, you tested soil for particle size and air content.
   (a) What other factors might make soil better or worse for growing plants?
   (b) (2E) Design an investigation to explore one of these factors. If your teacher approves your design, carry out the investigation.

Reflecting

3. How could you improve on the way you gathered and recorded observations in this investigation?

(2E) Designing an Inquiry Investigation

Soil and Plant Growth

Fertile soil is one of the world's most valuable resources. Soil that can support the growth of crops has always been in limited supply. In Canada, less than 1% of the total land area is prime agricultural land. For gardeners, good soil for both indoor and outdoor plants is essential. But what makes good soil?

Rich Soil Means Lots of Humus

Black or brown soil, rather than grey soil, has the most humus in it. It is the decaying plant and animal material in humus that gives soil its dark colour and supplies plants with most of the nutrients and minerals necessary for growth. Humus has a damp and sticky texture.

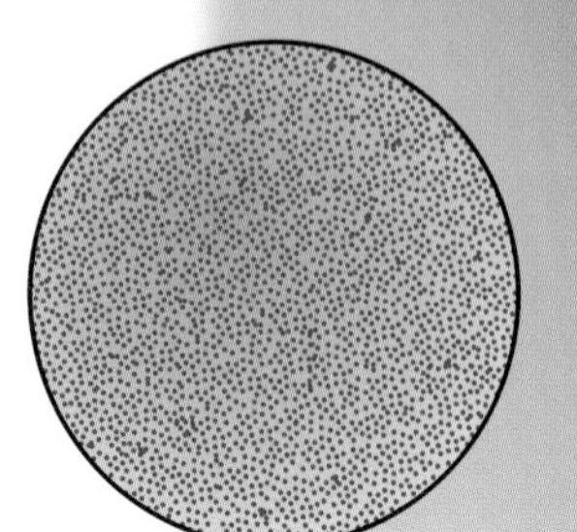

clay particles (less than 0.002 mm)

Fertile Soil Includes Sand, Silt, and Clay

Good soil includes roughly equal proportions of sand, silt, and clay. Each has a different texture because it is made up of particles of different sizes, as shown in **Figure 1**.

Between the soil particles are spaces. The spaces are important, because that is where the other ingredients of good soil can be found—water and air. Water and air are essential for plant growth. Plants absorb water with their roots. They use the water to make food. Bacteria and other organisms in the soil need air before they can break down humus into a form that plants can use. Minerals and other nutrients are also drawn into plants through their roots. Good farming soil contains roughly half particles and half spaces.

Because sand particles are large, the spaces between the particles are also large. These large spaces allow the tiny roots of plants to grow down through the soil more easily. However, the large spaces also allow water to drain away quickly, leaving plants with no water. Corn is just one plant that does not grow well in sandy soil because it requires a lot of water.

Because the spaces between clay particles are so small, clay holds water much better than sand. But clay particles also pack together tightly, preventing water from entering the soil easily. The small spaces between clay particles also make it difficult for plant roots to grow, and soil with lots of clay in it will have less air available to soil organisms.

Silt particles are in between sand and clay in size. They hold water better than sand, and they allow plant roots to grow better than clay. It's good to have some silt in soil, but soil made only of silt will not be as good as a mixture of all three types of particles.

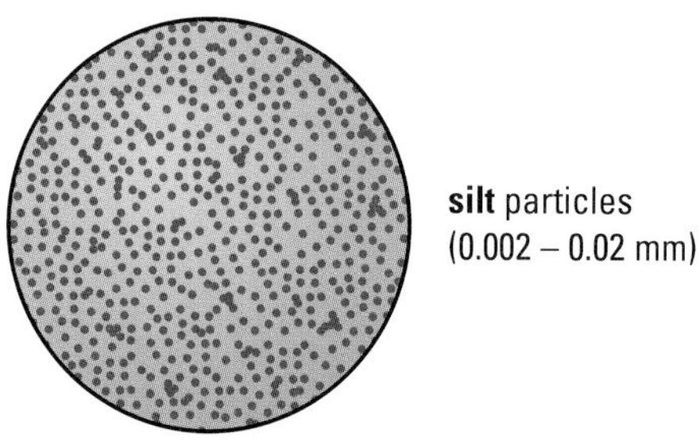

silt particles (0.002 – 0.02 mm)

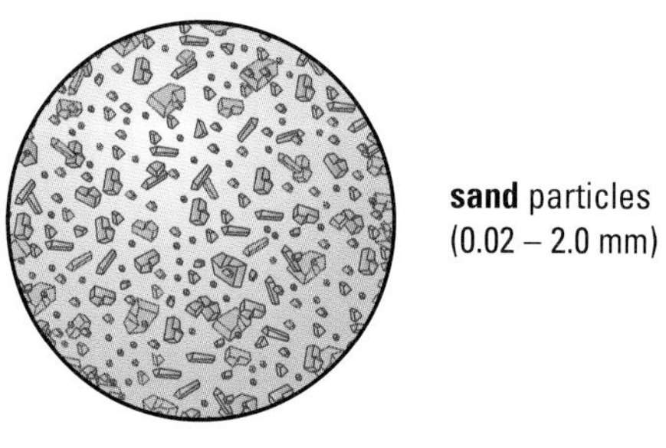

sand particles (0.02 – 2.0 mm)

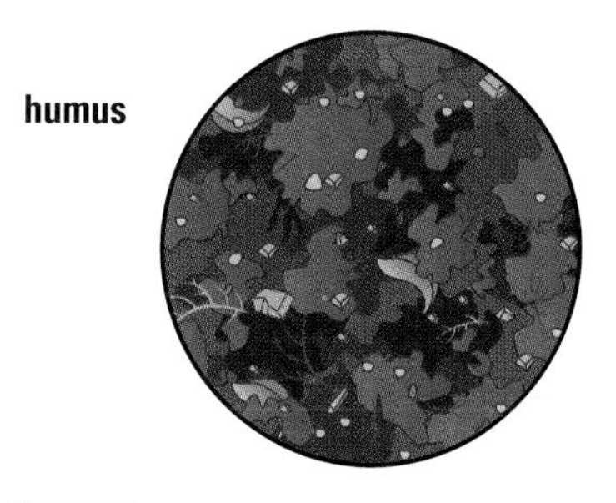

humus

Figure 1

Soil is made up of humus and particles of rock. The rock particles have different names, depending on how large they are.

Understanding Concepts

1. Why is the humus content in soil important?
2. Why are the spaces between soil particles so important?
3. Why does clay alone make a poor soil for growing most plants?

Making Connections

4. (a) On a baseball field after a rainstorm, puddles have formed around home plate and the pitcher's mound, and along the paths between the bases but nowhere else on the field. How can you explain this observation?
   (b) How could you improve the condition of the field?
5. Some houseplants, such as cacti, thrive in environments where there is very little water. What kind of soil mixture would be good for these plants? Why?

Design Challenge

When designing a tailings pond, what type of soil in the pond bed would best prevent the tailings from leaking into the environment?

Try This: Water-Holding Capacity of Soil

Soil particles affect how much water the soil can hold.

- Examine 3 different 50-mL samples of soil with a hand lens.

1. What types of particles can you observe in each sample? Describe the differences between the samples.
2. Predict which soil sample will hold water best.

- Place each soil sample in identical folded pieces of cheesecloth, and fasten the corners together with a rubber band, as shown in **Figure 2**.
- Measure 50 mL of water in a graduated cylinder.
- Hold the first cheesecloth sack over a pail.

Figure 2

- Slowly pour the water over the sack. Stop pouring when water begins to drip into the pail.
- Record the amount of water left over in the graduated cylinder.
- Repeat with the second sack.

3. Which soil sample held the most water? Was your prediction correct?

Farming and the Soil

All organisms, including humans, alter their environment simply by living in it. Farming is a dramatic example of a human effect on the environment. The way we grow food can change the soil, increase the chance that it will be eroded, and add chemicals to the environment.

Heavy Machinery

Heavy, mechanized farming equipment has greatly improved the efficiency of farmers in growing and harvesting crops, as shown in **Figure 1**. However, the frequent use of this heavy equipment presses down the soil, forcing soil particles closer together. The soil becomes compacted. Compacted soil has fewer and smaller spaces between the soil particles.

(a) How would soil compaction affect the soil's ability to grow crops?

Figure 1
Modern farms use heavy machinery to cultivate the soil and harvest crops.

Ploughing

Since the invention of the mechanized plough, farmers have been using it to **till**, or break up, the surface of the soil efficiently. During tilling, any hard crust on the soil is broken. This allows water and air to enter the upper layers of the soil. Planting seed becomes easier. Young plants grow more quickly in the loosened soil. But tilling also has a downside: it leaves the soil exposed to erosion.

Soil erosion results not only in the loss of valuable topsoil, but also in the removal of soil nutrients. **Zero tillage** (**Figure 2**) is one way of dealing with this problem. Using this method, stubble from the previous crop, including the roots, is left in the ground, and new seed is planted into the old stubble. No plough is used.

(b) Why would tilled land be more vulnerable to erosion than untilled land?

(c) How would stubble from the previous crop help to prevent erosion?

Figure 2
Planting in a zero-tillage field. This farmer has not ploughed the soil before planting.

Other Methods of Preventing Erosion

Some farmers plant rows of trees next to their farms and between fields. Others who have sloping land plant their crops in level rows that carefully follow the contours of the land, as shown in **Figure 3**, so that water does not run downhill along any row.

(d) Which type of erosion is prevented by trees?

(e) Which type of erosion is prevented by level rows of crops?

Figure 3
This field has been ploughed following the contours of the land to reduce erosion.

Changing the Crop

Although planting one crop in the same area every year is easier for the farmer, it does cause problems. Each crop removes different nutrients from the soil. For example, the nutrients used by the wheat plant to make its stalk and leaves may be returned to the soil when its stubble decays, but the nutrients in the wheat seeds (the grain) are taken by the farmer. If those nutrients are not replaced, the soil will become less and less productive. Farmers replace missing nutrients with fertilizer (see **Figure 4**).

There are other problems associated with repeatedly growing the same crop in the same field—pests and diseases. Insect pests, bacteria, and moulds that feed on that crop will stay in the field over winter and be ready to attack the next crop in the spring. Weeds that compete with the crop for the soil's nutrients will leave their seeds in the soil, ready for the next year. To deal with these problems farmers may resort to repeated use of insecticides, fungicides, and herbicides.

(f) Some farmers practise **crop rotation**, growing a different crop every year in the same field. How might crop rotation help farmers?

(g) Water runs off and through fields into neighbouring streams and rivers. What kinds of problems could the use of herbicides and insecticides cause downstream from the farm?

(h) Would growing two different crops in alternate rows reduce the farmer's need to add chemicals to a field? Explain.

Figure 4
Using the soil for the same crop every year forces farmers to add fertilizer to keep the soil fertile.

Understanding Concepts

1. Explain how the use of heavy machinery on farmland causes problems with the soil.
2. Name one problem related to modern farming methods and a possible solution.
3. What are the advantages of zero tillage?

Making Connections

4. Motorized machinery is a relatively new development for farmers. Its use has increased the production from most farmland, but there are problems involved in using it.
   (a) What disadvantages would there be to farmers who didn't use heavy machinery on their farms?
   (b) Is it likely that any new technology will bring only positive changes with it? Explain.

Design Challenge

Would any of the methods presented here help in the design of your field? How could you integrate them into your design?

Erosion: Carving the Landscape

Canada is a land of contrasts. In many places in the north, as in northern Ontario, the underlying rock of the Earth's crust, the bedrock, is exposed. Farther south, the bedrock is less visible, covered as it is in rich, thick soil. How can we explain this difference?

Ancient Ice Ages

Ancient glaciers are the surprising answer. Over the last 2 million years, five periods of glaciation have occurred in North America, creating huge changes in the features of the land. Each time, as the climate of Earth cooled, a thick ice sheet advanced slowly southward, covering much of North America. The most recent period of glaciation ended about 10 000 years ago. This last glacier, called the Wisconsin Glacier, covered North America from the Arctic to south of the Great Lakes, as shown in **Figure 1**.

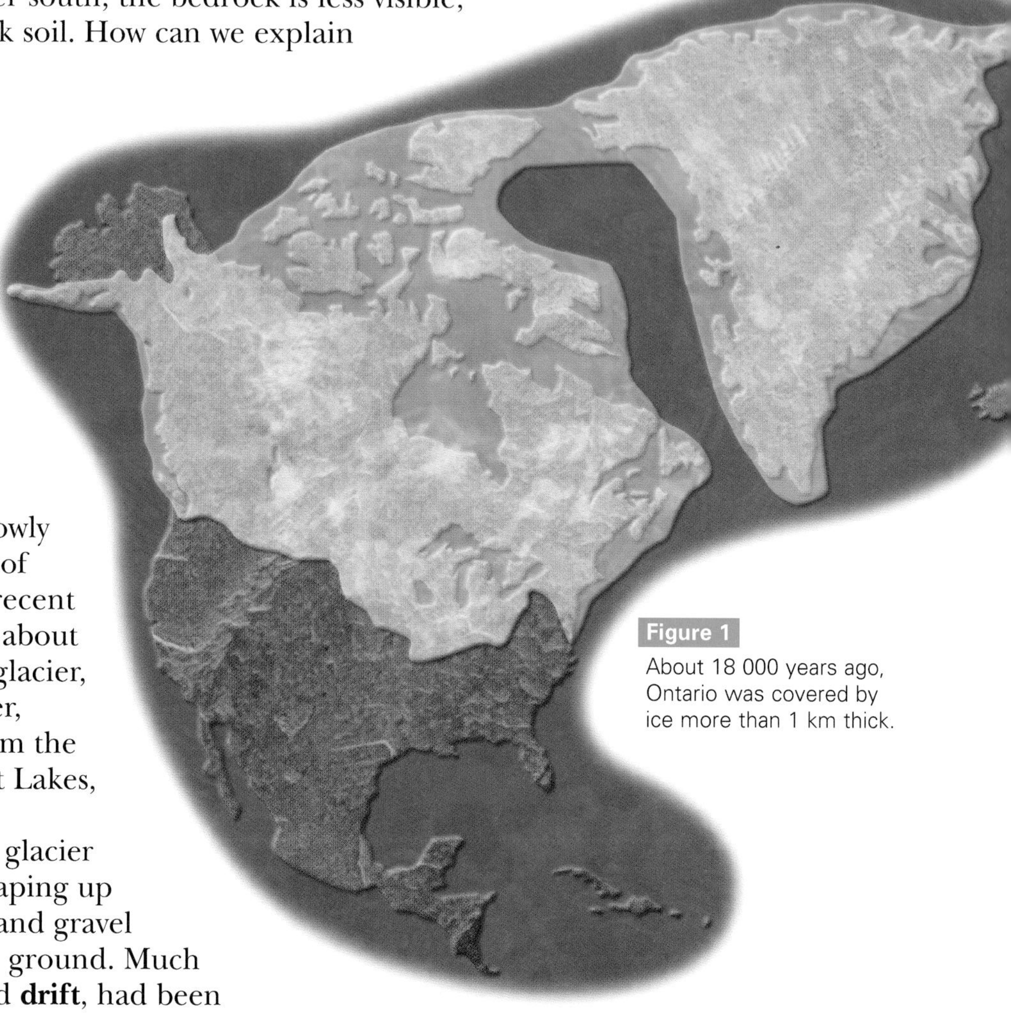

Figure 1
About 18 000 years ago, Ontario was covered by ice more than 1 km thick.

The great weight of the glacier caused it to spread out, scraping up bedrock, loose rock, sand, and gravel as it slowly pushed over the ground. Much of this loose material, called **drift**, had been left behind by previous glaciers. The glacial movement carried this material south and eventually deposited it in moraines, leaving the northern bedrock scraped clean. Many of the hilly areas north of Lake Ontario and Lake Erie are actually moraines, now covered with trees, fields, and towns.

Another feature of receded glaciers is called a **drumlin** (**Figure 2**). Drumlins look like narrow moraines, but are often in groups, all pointing in the same direction. Drumlins are caused by a glacier rearranging the moraine of a previous glacier, pushing the drift in the new direction.

Figure 2
Drumlins are narrow hills, usually in groups, formed from glacial drift. All point in the direction the glacier was moving.

How Ontario Looks Today

Many of the features that you see in Ontario's landscape can be traced to glacial action. Most of the small lakes, for example, were gouged from the bedrock by the glaciers. The Great Lakes' basins, too, were scraped larger and deeper with the passing of each glacier during the five ice ages. Ontario's many rivers drain the small lakes. They have eroded their channels through the layers of rock and gravel left behind by the glaciers.

Rivers and Flood Plains

Figure 3 shows how the erosive force of moving water has continued to shape the land since the last glacier receded. The first streams and rivers were fast flowing and fairly straight. Then, over thousands of years, they cut away at their banks, eroding material from the outside of curves, and depositing it at the inside of curves where the water moves more slowly. The curves eventually got larger and rounder, and the rivers became slower and winding. During the spring, when a river is full of meltwater from snow, it may overflow its banks and cover the shores with muddy water. The mud settles and helps build up and level out the **flood plain** on either side of the river.

Understanding Concepts

1. Why is there more soil in southern Ontario than in northern Ontario?
2. What erosive force created most of the lakes in Ontario?
3. Using a series of diagrams, explain how a river changes from being young and straight to old and winding.
4. Glaciers slow down the formation of soil on the one hand, yet speed it up on the other. Explain how this is true.

Exploring

5. A topographical map shows the change in height of the land, indicating the location of hills, lakes, and mountains. Obtain a topographical map of your area from the Geological Survey of Canada, and see if you can find evidence of old and young rivers, moraines, and drumlins left behind by glaciers.

Reflecting

6. There have been five ice ages in the last 2 million years. Do you think we have seen the last of the ice ages? Why or why not?

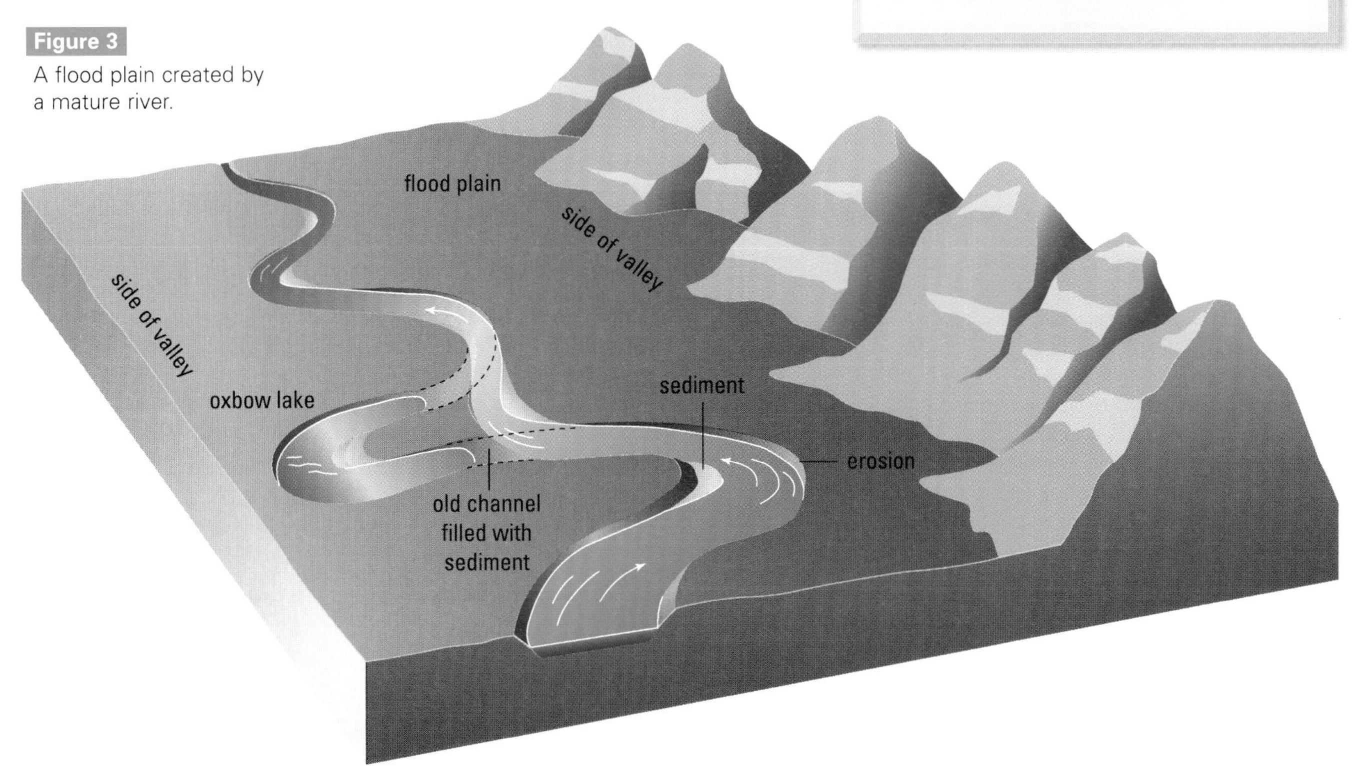

Figure 3
A flood plain created by a mature river.

4.12 Inquiry Investigation

SKILLS MENU
- Questioning
- Hypothesizing
- Planning
- Conducting
- Recording
- Analyzing
- Communicating

Mountains to Molehills

You have learned that many factors, including running water, cause rock to erode and that some rocks erode faster than others. Mountains are made of rock. What are the effects of rain, snow, ice, and streams on mountains over time? **Figure 1** shows the Rocky Mountains and the Laurentian Mountains. Water, wind, and ice continue to affect these mountains.

Materials
- apron
- paper
- potting soil
- clay
- sand
- popsicle sticks
- dishpan
- water
- watering can

Question

1 What question is being answered through this investigation?

Hypothesis

(2C) 2 Create a hypothesis for this investigation.

Experimental Design

In this investigation, you will design and build a mountain that will resist erosion. Erosion will be caused by a watering can full of water. Your design will influence how much your mountain erodes.

3 (6C) Design and draw a mountain shape that will use any of the materials available. Your mountain must be as resistant as possible to erosion. Include an explanation for the features of your mountain.

Procedure

4 In the dishpan, build your mountain according to your plan with the materials you have chosen.

5 Once you have finished building, examine the other mountains.

(a) Predict which mountain will best resist the erosion from the watering can. Explain your prediction.

6 One student will use the watering can to erode each mountain. Observe each mountain, including your own, as the water is poured onto it.

(a) How well did your mountain withstand the water? Draw a diagram showing your mountain after erosion.

SKILLS HANDBOOK: (2C) Predicting and Hypothesizing (6C) Scientific & Technical Drawing

Figure 1

Weathering and erosion work on (a) the Rocky Mountains, and (b) the Laurentian Mountains. What differences do you notice in the two ranges?

(a) The Rocky Mountains

Making Connections

1. As you have learned, erosion can also be a problem below ground since water seeps through soil and rocks. What features of the mountain you designed would help prevent this kind of erosion?
2. Look at **Figure 1**. The two photographs look very different.
   - **(a)** Make a T-chart and list some differences you notice between the two mountain ranges.
   - **(b)** How would you account for the differences?
   - **(c)** Predict what will happen to each mountain range over millions of years of erosion.

Design Challenge

Could you use any of the features you built into your mountain design in the design for a model field?

Analysis

7 Analyze your results by answering these questions.

(a) How good was your prediction? Which mountain stood up best to the watering-can test?

(b) From observations of your mountain and the others, which materials and features seem to work best at resisting erosion?

(c) Should you modify the hypothesis you wrote at the start of this investigation? Explain.

(d) If you had other materials available, what could you have done to prevent your mountain from eroding?

(b) The Laurentian Mountains

Mountains to Rock

You have learned how erosion breaks down mountains of rock. Streams and rivers carry all the soil and rock pieces, or **sediment**, downstream, as you can see in **Figure 1**. Where does all this sediment end up?

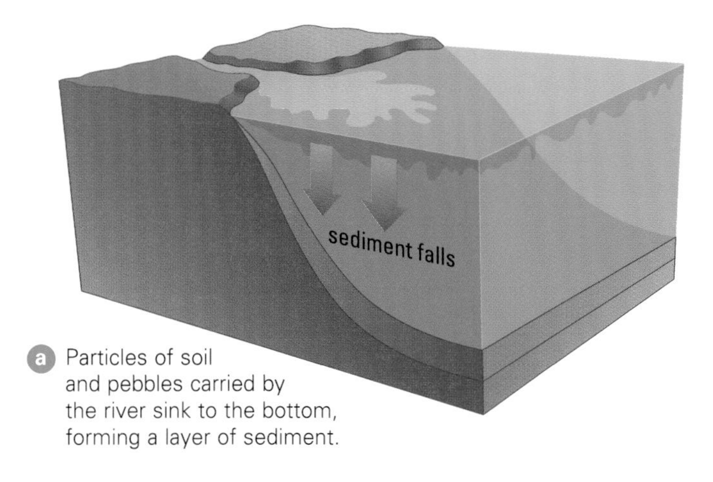

a Particles of soil and pebbles carried by the river sink to the bottom, forming a layer of sediment.

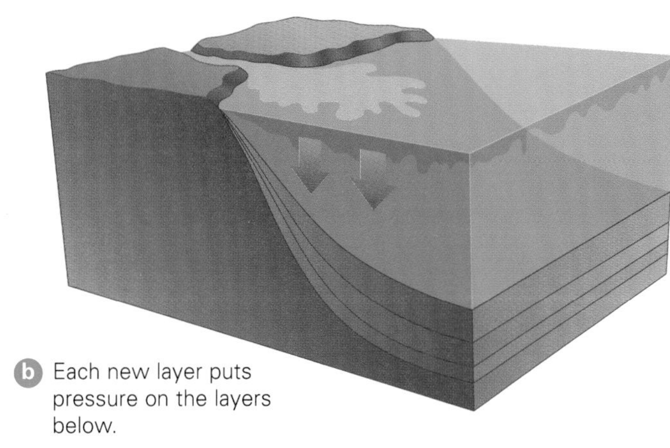

b Each new layer puts pressure on the layers below.

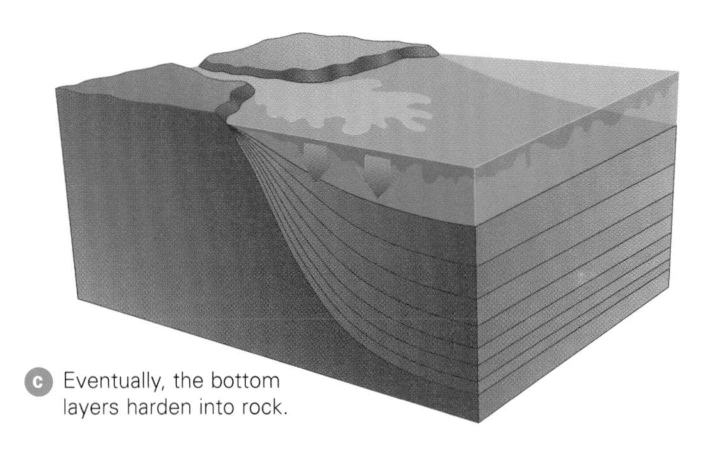

c Eventually, the bottom layers harden into rock.

Figure 2
Sedimentary rock is formed as layers of sediment are added by a river.

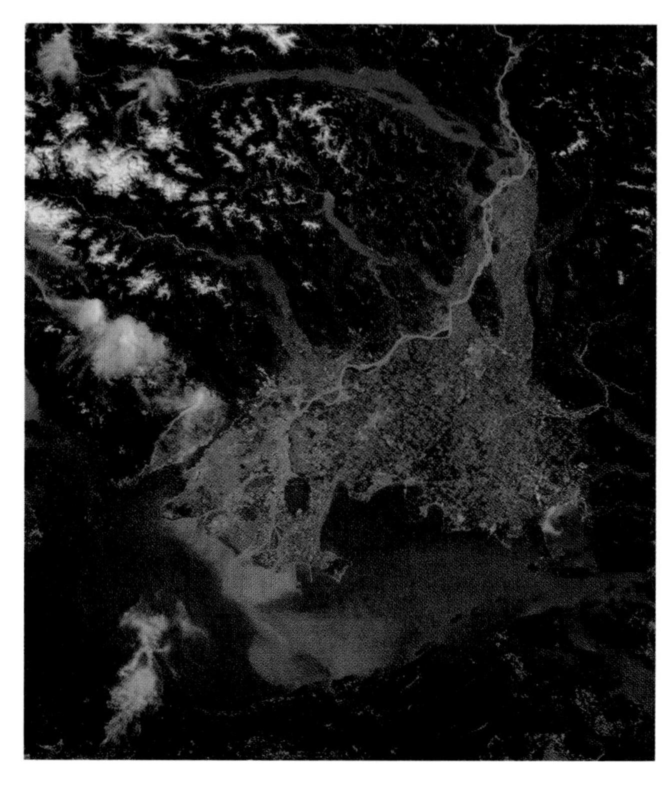

Figure 1
As this satellite photo shows, sediment carried by the Fraser River pours out into the sea.

Sediment Layers

The sediment is carried downhill by the strong current of small rivers and streams into larger rivers. Eventually, as the water approaches a lake or an ocean, the current slows, and the water can no longer carry the sediment along. The sediment gradually sinks to the bottom. There, on the lake or ocean floor, the sediment gradually piles up in layers. Over millions of years, the enormous weight of the many layers of sediment presses down on the lower layers. Under that pressure, water is squeezed out and the lower layers slowly harden into **sedimentary rock** (**Figure 2**).

Types of Sedimentary Rock

Different types of rock are formed from different types of sediment, as shown in **Figure 3**.

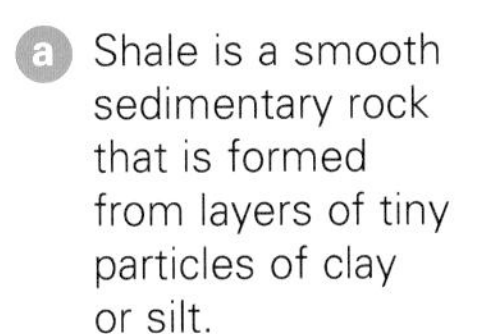

(a) Shale is a smooth sedimentary rock that is formed from layers of tiny particles of clay or silt.

(b) Sandstone, a rougher rock, is formed from layers of compressed sand.

(c) Conglomerate is made from sediment that contains pebbles and small stones.

Figure 3
As the layers of sediment are compressed into rock, they tend to form different kinds of rock, depending on the nature of the particles in the sediment.

Understanding Concepts

1. What part does erosion play in the formation of sedimentary rock?
2. Explain why there are different types of sedimentary rocks.
3. (6C) In a series of diagrams, show what eventually causes sedimentary layers to become hard.

Making Connections

4. The Niagara Gorge is being created by erosion caused by the Niagara River. What will happen to the rock eroded by the river?
5. In many parts of the world, including North America, there are areas of sedimentary rock that extend for hundreds of kilometres. What does this clue tell you about the ancient history of these areas?

Reflecting

6. The oldest rock in the world is about 4 billion years old, almost as old as Earth itself. However, this rock is not sedimentary. Speculate on how this rock might have formed.

Evidence of Sediment

There are many places we can see evidence of sedimentation and the formation of sedimentary rocks. The Niagara Escarpment, as shown in **Figure 4**, and the Grand Canyon, as shown in **Figure 5**, are just two examples.

Figure 4
The sedimentary layers that can be seen in the Niagara Escarpment were laid down millions of years ago in an ancient ocean bed that covered large parts of North America.

Figure 5
The Grand Canyon, which is over 1500 m deep, is being created as the Colorado River slowly erodes a path through layers of sedimentary rock laid down in an ancient ocean.

SKILLS HANDBOOK: (6C) Scientific & Technical Drawing

Fossils: Rock's Timekeepers

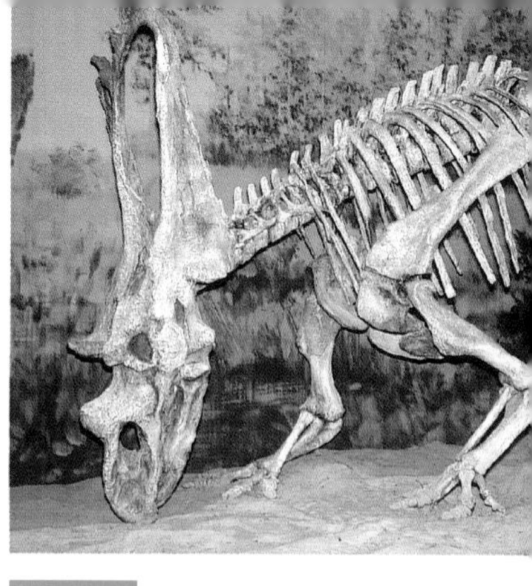

Figure 1
These fossil dinosaur bones, discovered in sedimentary rock in the Alberta Badlands, were reconstructed to form an entire skeleton. This animal lived about 75 million years ago.

You've probably heard of, if not seen, fossils of dinosaurs and other animals that lived on our planet millions of years ago. The dinosaur in **Figure 1** is an example. But what are fossils? How do they form?

How Fossils Form

Fossils are rocklike casts, impressions, or actual remains of organisms that were buried after they died, before they could decompose. Only a tiny fraction of organisms are preserved as fossils. This is because most dead organisms decay or are eaten by scavenging animals. Also soft tissue, such as muscle and other organs, does not fossilize well. Animals that have neither bones nor shells will not leave fossils.

An organism that is suddenly buried, for example if it falls into mud or quicksand, or is covered quickly by a landslide of sediment or blowing volcanic ash, may become a fossil. As the layer that contains the organism is covered by other layers of sediment, it gradually becomes sedimentary rock.

As wet sediment becomes rock, minerals that are dissolved in the water gradually replace minerals in the body of any buried organisms. Bone, shell, and the body parts of plants can all be replaced this way. Eventually, particle by particle, the fossilizing organism is replaced by minerals. The final result is a fossil that looks exactly like the original organism but is in a rocklike form (see **Figure 2**).

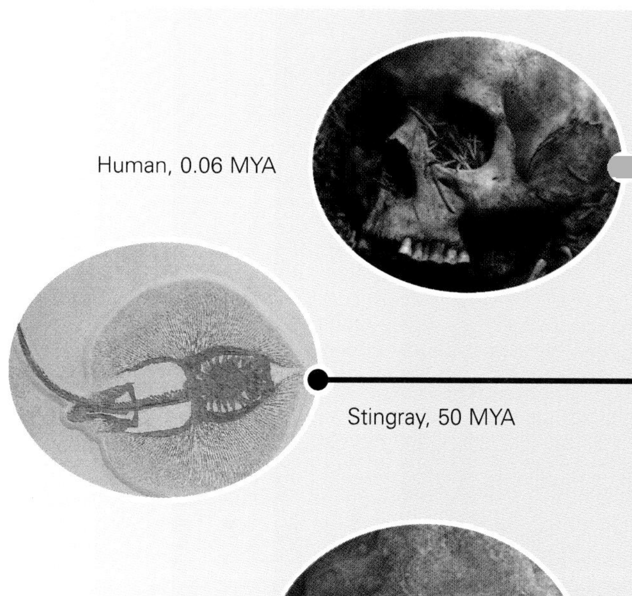

Nothosaur, 210 MYA

Tree fern, 300 MYA

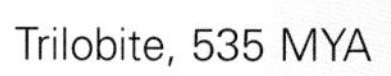

Figure 3
Geological time scale. MYA = millions of years ago

Figure 2
A fossilized log. Minerals have replaced the original wood, preserving the wood's structure. Entire forests, killed and preserved by volcanic eruptions or mudslides, have been fossilized in this way.

Fossils: Ancient Snapshots in Sedimentary Rock

Fossils are important because they show us what types of animals and plants lived on Earth hundreds of millions of years ago. If you look at exposed layers of sedimentary rock, from bottom to top, the fossils are like a series of snapshots of how life has changed on Earth, from the distant past near the bottom to more recent times near the top, as shown in **Figure 3**.

Fossils are also used to compare the age of rocks. For example, if a certain type of ancient fossil is found in two different places, then those two rock layers were probably formed at about the same time.

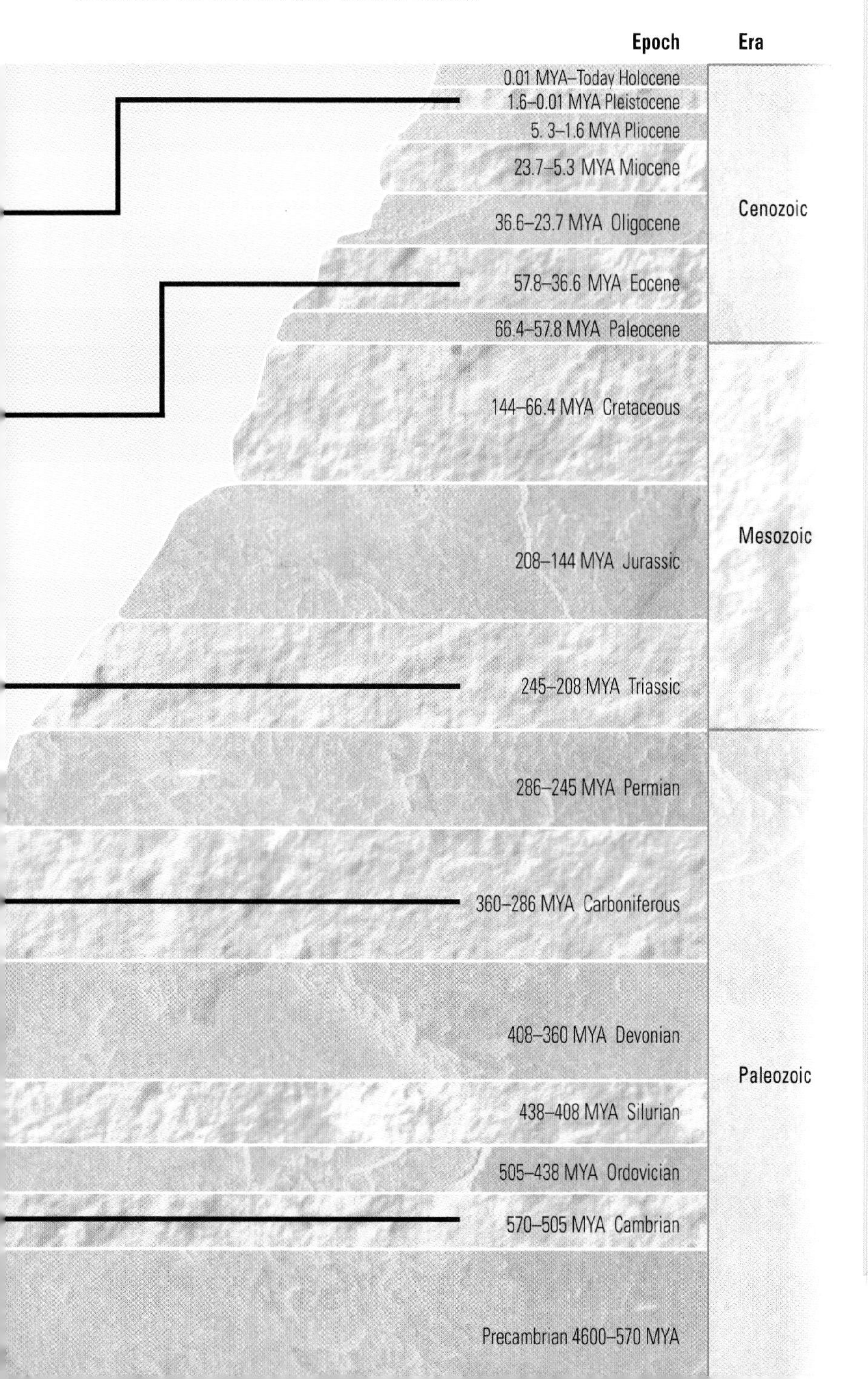

Understanding Concepts

1. What are fossils and how are they made?
2. Why are fossils rare?
3. If you found a rock with a fossil that looked like **Figure 4**, how old would the rock be?

Figure 4

Making Connections

4. No fossils have been found in the lowest sedimentary layer in the Grand Canyon. Does this mean that life did not exist on Earth when that layer of sediment formed? Explain.

Exploring

5. From fossil evidence, it appears that dinosaurs and many other animals died out suddenly about 65 million years ago. Some scientists think the mass extinction was caused by an asteroid that hit the Earth. Others believe massive volcanic eruptions were the cause. In both cases, the air would have filled with dust or ash, blocking the sun and cooling the Earth rapidly. What might have happened to a dinosaur caught in this disaster? Create a diagram showing the life, death, and fossilization of a dinosaur.
6. (4A) The Burgess Shale, near the town of Field in British Columbia, is one of the most famous fossil finds in the world. Research what types of fossils can be found there and create a brochure explaining why they are so important.

Reflecting

7. Based on what you have learned, why would scientists who have only fossil evidence have to be careful about interpreting what life was like on Earth?

4.15 Case Study

Drifting Continents

Many people have looked at the map of the world and noticed that some of the continents look like jigsaw pieces that have somehow become separated. Some came up with fanciful explanations, for example that missing continents had sunk into the sea. Among the explanations was the visionary hypothesis of Alfred Wegener, in the year 1912. He proposed that millions of years ago, all of the continents were actually part of one supercontinent he called Pangaea. The supercontinent had broken up and the continents had gradually moved apart to where they are now. He called his hypothesis **continental drift**.

At the time, most other scientists completely dismissed Wegener's idea because they could see no evidence to support it. Besides, they argued, how could huge continents move thousands of kilometres?

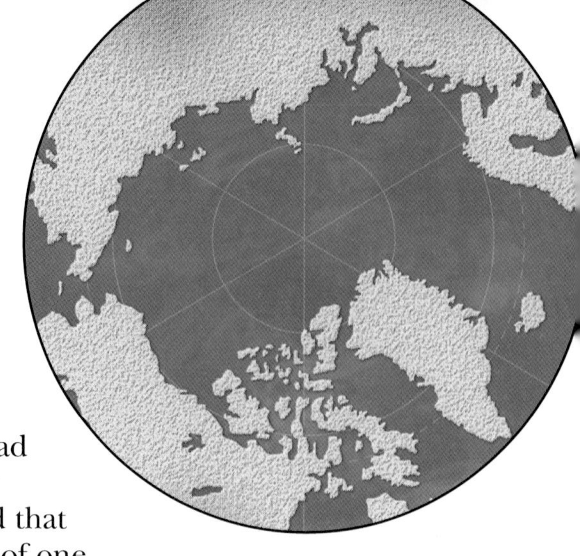

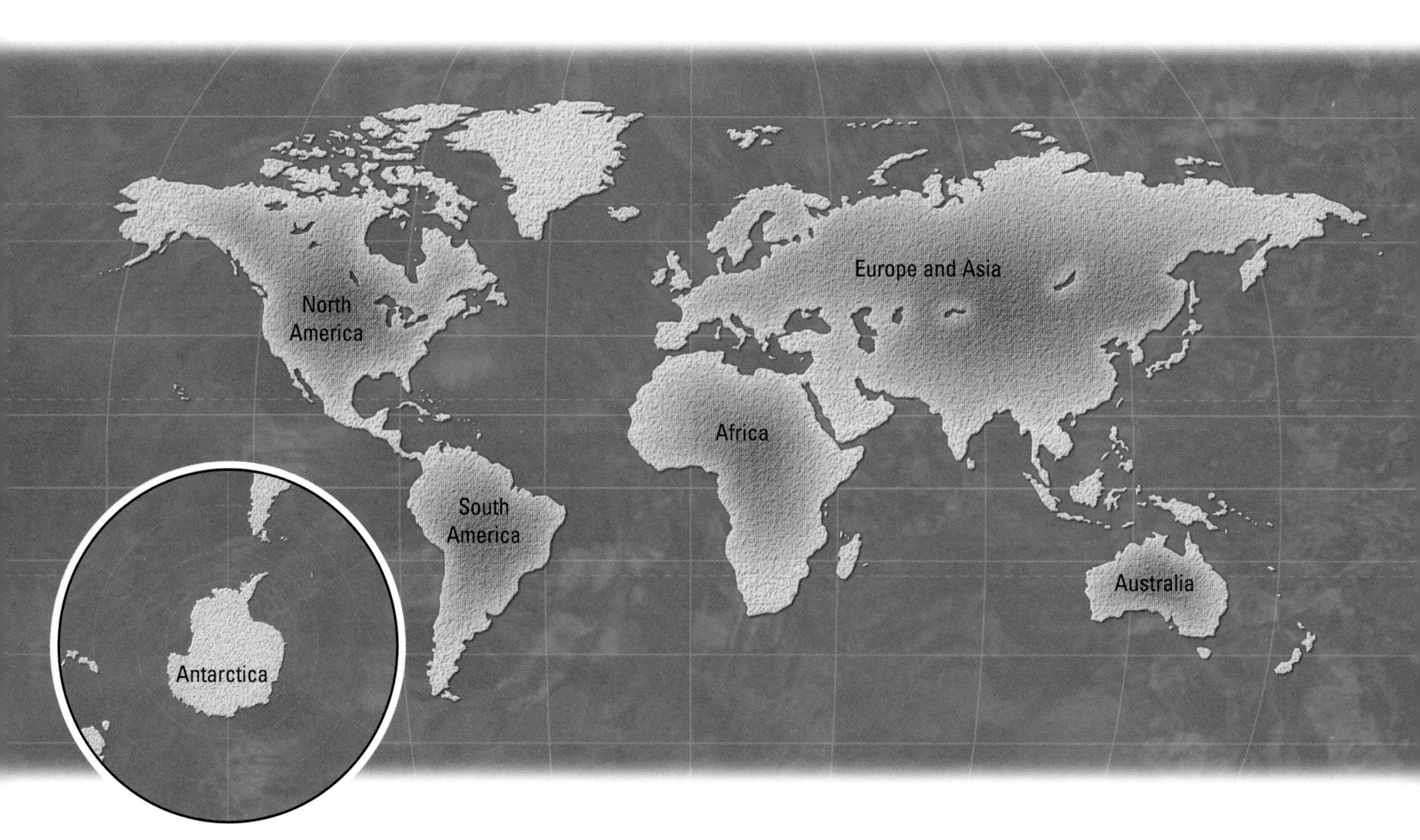

Figure 1
Were the continents once part of one giant supercontinent?

(a) Look at the continents in **Figure 1**. Do any of the continents look like they could fit together? Which ones?

(b) How might you explain why those continents seem to fit quite well but not perfectly?

(c) If you were a geologist, what features might you look for on the continents or in the rocks themselves that would help support or disprove continental drift?

The Search for Evidence

Once Wegener had put forward his hypothesis, geologists, biologists, and other scientists began to search the Earth for evidence that would support or disprove his idea. As the years passed, the supporting evidence became stronger and stronger.

(d) Ancient fossils of a small, extinct aquatic reptile have been found only in South America and Africa. Could this be evidence to support continental drift? Explain.

(e) Fossils of tropical trees and animals have been found in Antarctica. That continent is now almost completely covered in ice that is many kilometres thick. How can the presence of these fossils be explained? Is there more than one explanation?

(f) Marsupials are unusual animals. Their babies are born like those of other mammals but in an earlier stage of development. The babies crawl into a pouch in their mother's body, where they continue to develop. Most marsupials live in Australia, but not all. Opossums are marsupials, and opossums live in North and South America. Does the existence of opossums support continental drift? Explain.

Plate Tectonics

Wegener's hypothesis is now widely accepted and is part of an established theory called **plate tectonics**. The theory of plate tectonics is that the Earth's crust is actually made up of several large sections, called **plates**, that are always moving slowly. These plates are floating on the hot, thick mantle kilometres below. The continents move by riding piggyback on top of these vast, thick plates of rock.

(g) Does the existence of earthquakes support the theory of plate tectonics? Speculate on how moving plates might cause earthquakes.

Understanding Concepts

1. Describe the theory of plate tectonics.
2. What clues are there that continental drift has been occurring for millions of years?

Making Connections

3. A large deposit of gold is discovered on the west coast of Africa. According to the theory of plate tectonics, where else in the world might a similar deposit be found? Use a map to support your opinion.
4. You can create a model of the Earth's crust using a hard-boiled egg. Lightly crack the shell of a hard-boiled egg, and use a felt-tip pen to outline the cracks. Remove a few pieces of the shell.
   - **(a)** Try sliding the remaining pieces around the surface of the egg. How is the eggshell a model of the Earth's crust?
   - **(b)** What happens when you slide pieces of eggshell into one another? Does anything similar happen to the Earth's crust?
   - **(c)** What activity at the Earth's surface cannot be modelled using an eggshell?

Reflecting

5. Wegener was laughed at when he first told other scientists about his hypothesis. A huge amount of evidence from every area of science was required before continental drift and the plate tectonic model that explained it were finally accepted. All new scientific theories must have a lot of evidence to support them. Are there drawbacks to this approach?

Moving Plates

The Earth's crust is not one continuous piece but is actually composed of several huge, solid sections, called plates (see **Figure 1**), that move slowly as they float on the semi-liquid mantle below. The plates are moving relative to one another. For example, the plate that carries North America and the plate that carries Europe are moving away from each other at the rate of about 3 cm every year. Some plates are moving toward each other. Some are slipping by each other. Wherever plates meet, earthquakes signal their movement.

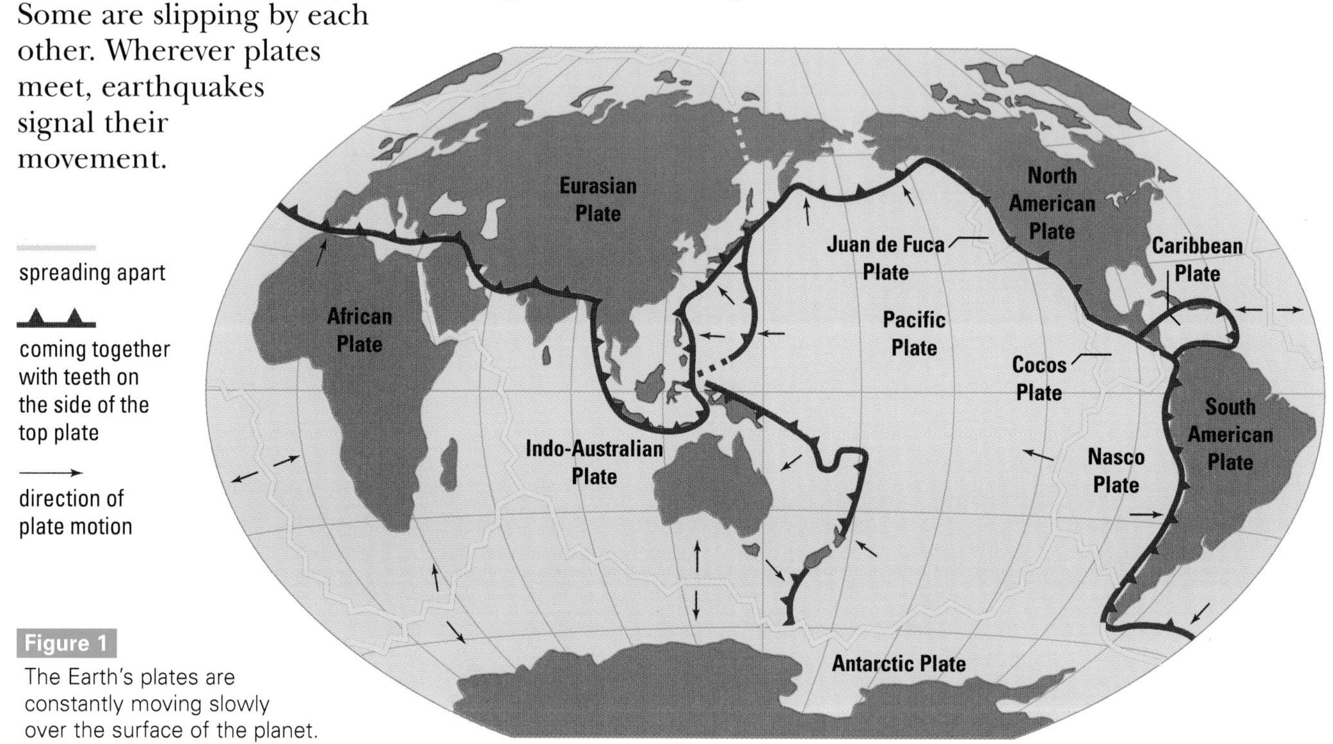

Figure 1
The Earth's plates are constantly moving slowly over the surface of the planet.

Slipping By: A Fault

The San Andreas Fault in California (see **Figure 2**) is an example of a place on the Earth's crust where plates are slipping by each other. Here the Pacific Plate, which carries part of Baja California and a small strip of the west coast of the United States, is moving north past the North American Plate. Places where plates meet in this way are called **faults**.

Figure 2
The famous San Andreas Fault in California. Here the Pacific and North American Plates meet. The Pacific Plate is moving north, and the North American Plate is moving south.

Collision: Subduction

Plates can collide as they move toward each other. When they do, something must give. Usually one plate slides below the other (see **Figure 3**). As the lower plate plunges underneath, it pushes into the hot mantle, where it heats up and melts. This process is called **subduction**. It is occurring in many places in the world, including along the coast of British Columbia.

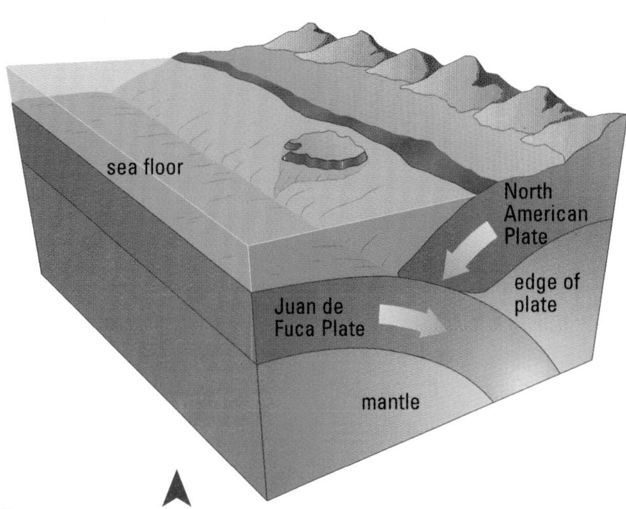

Figure 3
The Juan de Fuca Plate has been forced under the North American Plate, which is moving west.

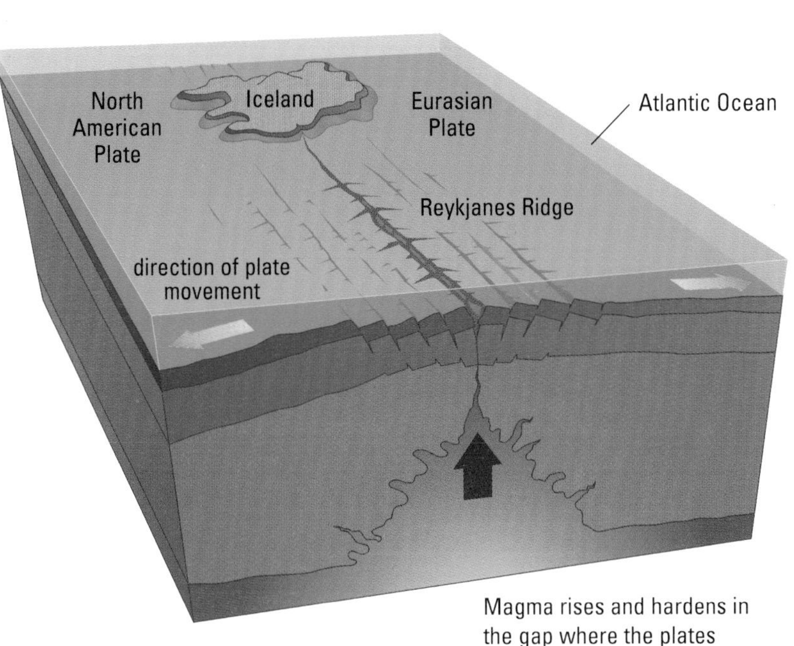

Figure 4
Iceland formed on the rift in the middle of the Atlantic Ocean where two large plates are moving apart. Magma from below continually rises up into the newly formed gap and flows out onto the Earth's surface as lava.

Separation: Ridges

There are also places on Earth where plates are moving apart from each other. At these spots, hot **magma** rises up into the crack between the plates and hardens, forming **ridges** of new rock (see **Figure 4**). Ridge formation usually occurs beneath the oceans, but one place on the Earth's surface where you can actually see this happening is in Iceland (see **Figure 5**).

Figure 5
In 1963, the new island of Surtsey formed off the coast of Iceland as large amounts of magma flowed out of the oceanic rift between the Eurasian and North American Plates.

Understanding Concepts

1. What are plates?
2. **(a)** Describe what happens when two plates meet and when two plates move apart.
   (b) Design a simulation of plates (3) meeting.

Making Connections

3. Look at **Figure 1**. Why might Vancouver have more earthquakes than Toronto or Halifax?

Exploring

4. Where on the Earth would you expect to find young, newly formed rock? Where would you expect to find old rock? Explain.
5. (4A) In Iceland, where hot magma is close to the surface, many people use its heat, called geothermal energy, to warm their homes. Research this form of energy. Is it renewable? Would it be practical to use geothermal energy in your area? Prepare a report.

SKILLS HANDBOOK: (3) Process of Design (4A) Research Skills

4.17 Inquiry Investigation

SKILLS MENU
- ○ Questioning
- ● Hypothesizing
- ○ Planning
- ● Conducting
- ● Recording
- ● Analyzing
- ● Communicating

Earthquakes, Volcanoes, and Mountain Ranges

The plates that make up the Earth's crust are constantly moving. They are also pushing against one another. Is there a direct connection between plate movement and earthquakes? Are mountain chains and volcanoes related to plate movement?

Question

Is there a pattern in the location of earthquakes, volcanoes, and mountain ranges (**Figure 1**)?

Hypothesis

1 Based on your knowledge of plate tectonics, create a hypothesis for each of the following questions. (2C)

(a) Where are earthquakes most likely to occur?

(b) Where are volcanoes most likely to occur?

(c) Where are mountain ranges most likely to occur?

Experimental Design

On a map of the world, you will plot the location of mountain ranges and recent earthquakes and volcanic eruptions to see if any patterns exist.

Procedure

2 Practise finding latitude and longitude on a map.
- To find latitude, measure from 0° at the equator up to 90° north at the North Pole or 90° south at the South Pole.
- To find longitude, measure from 0° at Greenwich, England, either east or west up to 180°.
- As a trial example, find the city in Ontario that is located at 46° N latitude and 76° W longitude.

3 On your copy of the world map, place a small blue circle ● at the location of each earthquake listed in **Table 1**.

4 For each volcano listed in **Table 2**, place a small red triangle ▲ on your world map.

5 Use a third colour to mark the location of the following mountain ranges: Rockies, Andes, Himalayas, Alps, Urals, and Appalachians. (You can use an atlas or globe to help you find these features.)

Materials

- map of the world
- atlas or globe
- coloured markers

Analysis

6 Analyze your results by answering these questions.

(a) Compare your map with Figure 1 in 4.16. What patterns do you see in the locations of:

- earthquakes?
- volcanoes?
- mountain ranges?

(b) Are volcanoes always located near mountain ranges?

(c) Were your predictions correct? If not, explain why.

SKILLS HANDBOOK: (2C) Predicting and Hypothesizing

(a) The 1989 earthquake in San Francisco, California was devastating.

Making Connections

1. Explain, using evidence from your map, why the edge of the Pacific Ocean is often called the Ring of Fire.

Design Challenge

Earthquakes are dangerous to miners. Could an underground mine be made safe during an earthquake? Would a tailings pond contain its waste during an earthquake?
How would you simulate an earthquake in your models?
What extra precautions would you have to take if your mine were on the west coast of Canada?

(b) Kilauea Iki, in Hawaii erupted in 1959.

Figure 1
Some areas of our planet seem to be more dangerous than others.

Table 1 Major Earthquakes

Year	Location	Coordinates
1556	Shenchi, China	39° N, 112° E
1755	Lisbon, Portugal	39° N, 9° W
1811–12	New Madrid, Missouri	36° N, 89° W
1906	San Francisco, Calif.	38° N, 122° W
1920	Kansu, China	40° N, 75° E
1923	Tokyo, Japan	36° N, 140° E
1935	Quetta, Pakistan	30° N, 67° E
1939	Concepcion, Chile	37° S, 73° W
1964	Anchorage, Alaska	60° N, 150° W
1970	Yungay, Peru	9° S, 78° W
1972	Managua, Nicaragua	12° N, 86° W
1976	Guatemala City	14° N, 91° W
1976	Tangshan, China	40° N, 119° E
1985	Mexico City, Mexico	19° N, 99° W
1988	Shirokamud, Armenia	41° N, 44° E
1989	San Francisco Bay, Calif.	38° N, 122° W
1990	Rasht, Iran	37° N, 49° E
1991	Valla de la Estrella, Costa Rica	10° N, 84° W
1993	Maharashtra, India	23° N, 75° E
1994	Northridge, Calif.	34° N, 119° W
1995	Kobe, Japan	34° N, 135° E

Table 2 Some Active Volcanoes

Volcano and Location	Coordinates
Etna, Italy	37° N, 15° E
Tambora, Indonesia	8° S, 117° E
Krakatoa, Indonesia	6° S, 105° E
Pelée, Martinique	14° N, 61° W
Vesuvius, Italy	41° N, 14° E
Lassen, California	40° N, 121° W
Mauna Loa, Hawaii	21° N, 157° W
Paricutin, Mexico	19° N, 103° W
Surtsey, Iceland	63° N, 20° W
Kelud, Indonesia	8° S, 112° E
Arenal, Costa Rica	10° N, 84° W
Eldfell, Iceland	65° N, 23° W
Mount St. Helens, Wash.	46° N, 122° W
Laki-Fogrufjoll, Iceland	64° N, 18° W
Kilauea, Hawaii	22° N, 159° W
Mount Katmai, Alaska	58° N, 155° W
Avachinsky, Russia	53° N, 159° W
El Chichon, Mexico	17° N, 93° W
Ubinas, Peru	16° S, 71° W
Villarica, Chile	39° S, 72° W
Asama, Japan	36° N, 138° E
Shikotsu, Japan	41° N, 141° E

4.18 Career Profile

Cracking the Secrets of the Earth's Crust

Meet Michael Schmidt. He is on "earthquake watch." He keeps an eye on the shifting continental plates of Canada's biggest earthquake zone.

People have always used stars for navigation. Michael Schmidt uses artificial stars that are 20 000 km above us. They're satellites that make up the global positioning system, or GPS. With GPS, Michael can pinpoint the slow creep of huge continental plates.

"The concept of how it works is very simple," says Michael. The GPS satellites send radio signals to Earth. Each signal includes a message that says exactly when it was sent. It's accurate to a billionth of a second! On Earth, GPS receivers pick up the signal and record when the signal arrives. The difference between sending and receiving is the signal's travel time. The signal moves with the speed of light, a number we know. The GPS computers multiply the speed of light by the signal's travel time to give the distance between the satellite and receiver.

This same multiplication is done with signals from at least four separate satellites. This gives the exact location of the receiver on the Earth.

Michael uses receivers that are attached to concrete pillars driven into the crust. When a continental plate moves, the receivers move with it (**see Figure 1**).

Figure 1
Michael Schimdt placed this temporary GPS receiver near the top of Mount Logan in B.C.

A Sticky Situation on the West Coast

Michael is most concerned about the North American and Juan de Fuca Plates on Canada's west coast. Scientists believe the Juan de Fuca Plate started to slip under the North American Plate, but, at the moment, the plates are stuck, as shown in **Figure 2**. Michael measures the movement of the western edge of the huge North American Plate as it is being squeezed out of shape or deformed. Each year the underlying Juan de Fuca Plate is pushing the North American Plate up about 4 mm and backwards (east) about 1 cm.

Added up over hundreds of years, these tiny movements store a huge amount of energy, like a tennis ball being squeezed tightly. When the plates unlock some day, the bulging North American Plate will suddenly leap forward over the Juan de Fuca Plate. "That's an incredible force happening all at once," says Michael. A tremendous earthquake will rock the coast.

Predicting the "Big One"

"GPS helps tremendously," says Michael. "If an earthquake like this happens under Vancouver, it's going to have an incredible effect."

Scientists believe the earthquake will happen where the most deformation takes place. Michael's GPS research shows that that spot is under the ocean about 120 to 150 km off Vancouver Island. Knowing this, Michael and his fellow scientists will be able to predict how strong the shock will be in nearby cities and towns. Using that information, architects and engineers may be able to make sure schools and other buildings are earthquake-proof and that will save lives.

Try This Unsticking Plates

You can demonstrate the movements of the continental plates that Michael observes using a sheet of paper, a sheet of heavy cardboard, and a ruler.

- Label the paper "North American Plate" and the cardboard "Juan de Fuca Plate."
- Place the cardboard and paper side by side, with the paper on the right. Put your right hand on the right half of the paper to hold it still.
- With your left hand, push the cardboard to the right against the edge of the paper. The left portion of the paper sheet should bulge up as it is squeezed. Keep pushing until the paper slips over the cardboard.
- Repeat this exercise with a partner. Let your partner slowly push the cardboard, then measure the distance the cardboard moved underneath the paper before the paper regained its shape.
- Start again, but this time measure the maximum height of the paper bulge before it slips.

1. Explain what might happen with real continental plates if this sudden movement took place next to the ocean.
2. How might this sudden motion affect living things?
3. Do you think your model of plate movement accurately represents what is occurring (or will occur) with the real plates on the coast of British Columbia? What are the weaknesses of this model?

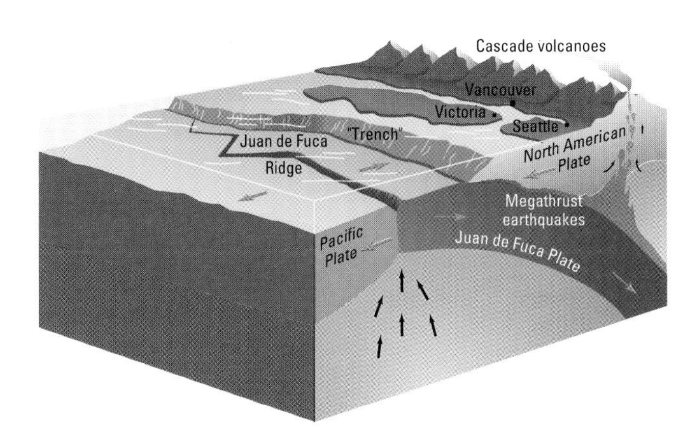

Figure 2
When the plates unlock, the "big one" will be let loose.

4.19 Design Investigation

SKILLS MENU
- Identify a Problem
- Planning
- Building
- Testing
- Recording
- Evaluating
- Communicating

Preparing for Earthquakes

If an earthquake happens near a city, it can be devastating, as **Figure 1** shows. During an earthquake, people are most in danger from collapsing buildings and falling pieces of concrete, stone, or steel. So the most effective way to minimize damage from an earthquake is to construct better buildings. Even in eastern Canada, where earthquakes are uncommon and usually mild, engineers must design office buildings to withstand at least small earthquakes.

Materials

- 1 m of masking tape
- 3 balls of modelling clay
- 100 toothpicks
- 30 thin wood sticks, each 4 cm long
- 2 thin pieces of wood, each 10 cm × 10 cm

The shaker must be careful not to damage desks or injure nearby students.

Problem

(3B) **1** What problem can you identify in the paragraph above?

Design Brief

Design and make an earthquake-proof building using only the materials available.

Design Criteria

- The building must be at least 30 cm tall and not more than 20 cm wide.
- The building must be taped to the desk on which it is built. A building that comes loose will be judged to have collapsed.

Build

2 (3D) Design at least two earthquake-proof buildings.

3 (3C) Pick the design you think is better, and build it.

(a) How did you decide which design is the better? What are the important features of the design?

4 (3E) When your model has been built, present it to the rest of the class, pointing out the features of your design that you think make your model earthquake-proof.

Test

5 (3F) Study each of the other models.

(a) Predict what will happen to each structure when the desk it is sitting on is shaken. Record your prediction.

6 One person will be chosen to shake the desk under each model.

- Starting with a very gentle shake, increase the amount of shaking every 10 s until the model falls down or the shaker gives up.

(b) Record your observations for each model.

SKILLS HANDBOOK: (3) Process of Design

Figure 1

The powerful earthquake that struck Kobe, Japan, in 1995 lasted only 20 s, yet it left 5502 people dead and $99 billion in property damage.

Evaluate

7 Evaluate your results by answering these questions.

(a) Based on the results of the testing, create a new design for an earthquake-proof building. Label the features of your new building and explain their purpose.

(b) Are the models you designed good examples of what real buildings might be like? Why or why not?

Making Connections

1. Why don't all new buildings contain features that will help them withstand earthquakes?

2. Look around at the buildings, bridges, and other large structures in your town or city.

(a) Can you see any evidence of reinforcements for earthquakes?

(b) Are these reinforcements just for earthquakes? If not, list other possible functions they might have.

Exploring

3. (4A) The CN Tower in Toronto is the highest structure in Canada. Research what features it has to protect it from earthquake damage.

Reflecting

4. No one yet knows how to predict earthquakes, but there is some evidence that some animals can. For example, rats and mice have been observed running from buildings just before an earthquake.

(a) What do you think these animals might be sensing?

(b) How could scientists test the reliability of these observations?

Design Challenge

You have tested several designs for a structure that resists earthquakes. Can you use any of the design features you created in your mine model? In your tailings pond model?

(4A) Research Skills

4.20

Mountains from Rocks

You have learned that mountains seem to be associated with the edges of plates. They form where two of the Earth's plates meet.

Fold Mountains

Where plates meet head-on, as the lower plate plunges into the mantle, the crust on the upper plate may fold under the pressure, forming **fold mountains** like the Rocky Mountains of North America and the Himalayas of Asia.

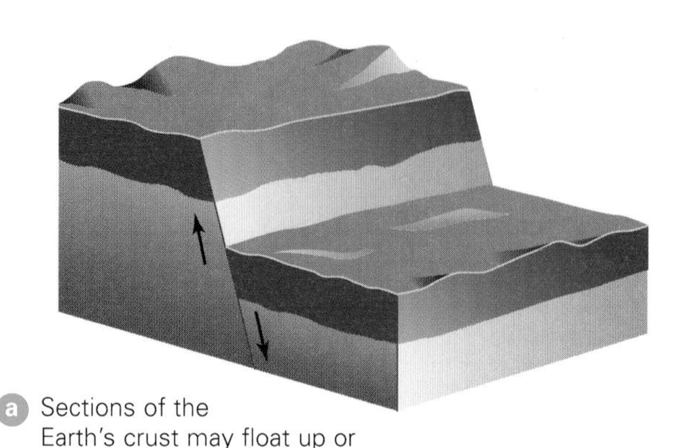

a Sections of the Earth's crust may float up or sink down past each other.

The Coast Mountains

The west coast of North America is mountainous. This is because the North American plate is moving generally west, colliding with the plates under the Pacific Ocean. The Juan de Fuca Plate under the Pacific Ocean at British Columbia is being forced down under the plate carrying the continent. As you can see in **Figure 1**, the tremendous pressure created by the lower plate causes mountains to form on the upper plate. Over millions of years, this continuous folding into high ridges and deep valleys has formed the Coast Mountains and the Rockies in North America.

b A block of the Earth's crust may sink between two fault lines, forming a huge valley, or it may rise, forming a large mountain.

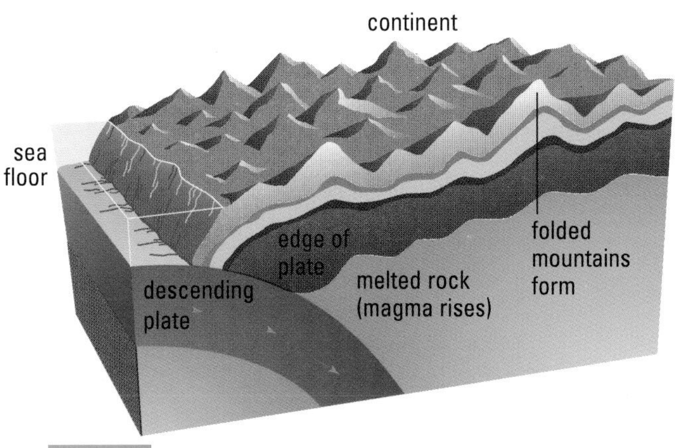

Figure 1

At the British Columbia coast, the North American Plate is sliding over the Juan de Fuca Plate. The huge force of the lower plate pressing against the upper plate as it thrusts downward causes the upper plate to rise, bend, and fold.

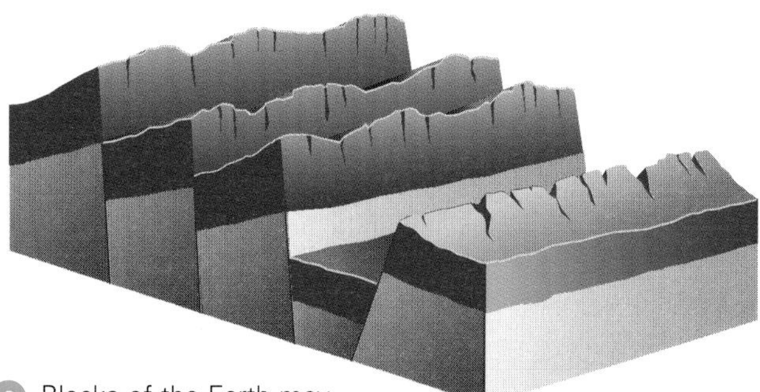

c Blocks of the Earth may move upward and tilt at the same time. The Sierra Nevada Mountains and the San Andreas Mountains in California are examples of block mountains that formed as blocks of the Earth's crust tilted and rose along two fault lines.

Figure 2

Block mountains or deep valleys may form along faults in the Earth's crust.

The Himalayas

The mountains in the highest range in the world, the Himalayas, are also fold mountains that have formed from two plates colliding. The plate carrying India is moving north, colliding with the large Eurasian Plate in front of it, and is plunging beneath it. Mount Everest, the highest mountain in the world, is getting still higher, by about 6 mm each year.

Mountains from Faults

Folding of plates during head-on collisions is not the only way mountains can be created. You have learned that plates move away from each other and toward each other, and they also slip past each other in areas called faults. However, plates are not smooth. As they slip by each other, they tend to have a grinding stop-and-start movement as the edges of the plates catch and push on each other. The pushing generates great pressure in the plates and can cause cracking as pieces of one plate or the other are forced up or down. The results of movement along a fault can be seen in **Figure 2**. **Block mountains** form as the crust tilts under pressure from a neighbouring plate.

Mountains from Below

A third type of mountain, not nearly as high as fold or block mountains, is called a **dome mountain**. Dome mountains form when magma moving up through the mantle encounters sedimentary rock in the Earth's crust that will not crack. The magma then simply pushes the layers above into a dome that rises higher than the surrounding land (see **Figure 3**).

Mount Royal, in the centre of Montreal, is a dome mountain. In the countryside surrounding Montreal, there are several other dome mountains. They rise only a few hundred metres, but they stand out prominently against the flat background.

Understanding Concepts

1. Use diagrams to explain the formation of:
   (a) fold mountains
   (b) block mountains
   (c) dome mountains

Making Connections

2. Can earthquakes be linked to the formation of all mountains? Explain.

Exploring

3. (a) Research how geologists estimate the age of mountains. (4A)
   (b) Which mountains are older, the Rocky Mountains in western Canada or the Laurentian Mountains in Quebec?

Reflecting

4. (8A) People have often considered burying toxic or radioactive waste far underground, where they hope it won't harm the environment. Based on your knowledge of plate tectonics, mountain formation, and earthquakes, is there a place where it would be safe to bury toxic waste? Prepare a report to support your opinion.

Figure 3
Formation of dome mountains

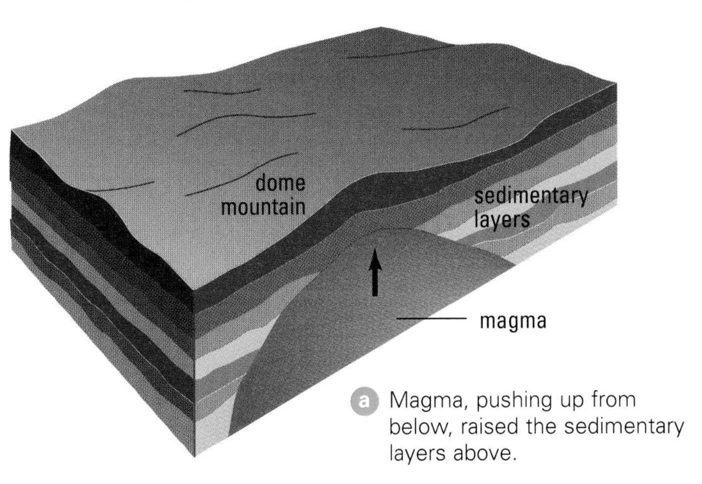

(a) Magma, pushing up from below, raised the sedimentary layers above.

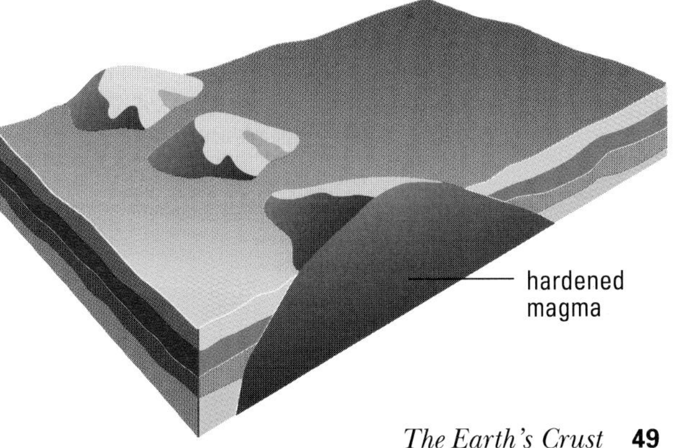

(b) As glaciers advanced and receded across the mountain, they eroded away much of the sedimentary rock, leaving the hardened magma exposed.

SKILLS HANDBOOK: (4A) Research Skills (8A) Writing a Report

Igneous and Metamorphic Rocks

About 4 billion years ago, the Earth's crust formed when the liquid magma on the surface cooled and hardened. Rock that forms from the hardening of liquid magma is called **igneous rock**. Four billion years ago, all of Earth's rock was igneous. Since then sedimentary rocks have formed, but there are also other types of rock formed by processes in the Earth's crust.

Igneous Rock

Most of the world's rock is still igneous, and igneous rock is still being formed.

Just below the Earth's surface, slowly cooling magma forms **intrusive igneous rock** that can be seen only after erosion removes the layers above it. Granite forms in this way.

Extrusive igneous rock is formed when magma (lava) pours out of volcanoes and hardens. Basalt is an extrusive igneous rock. It is the most common rock in the Earth's crust.

Metamorphic Rock

Below the Earth's surface, where rock is exposed to high heat and pressure, igneous rock and sedimentary rock form new types of rock called **metamorphic rock**. **Figure 1** shows how all of the rocks mentioned in this section are formed.

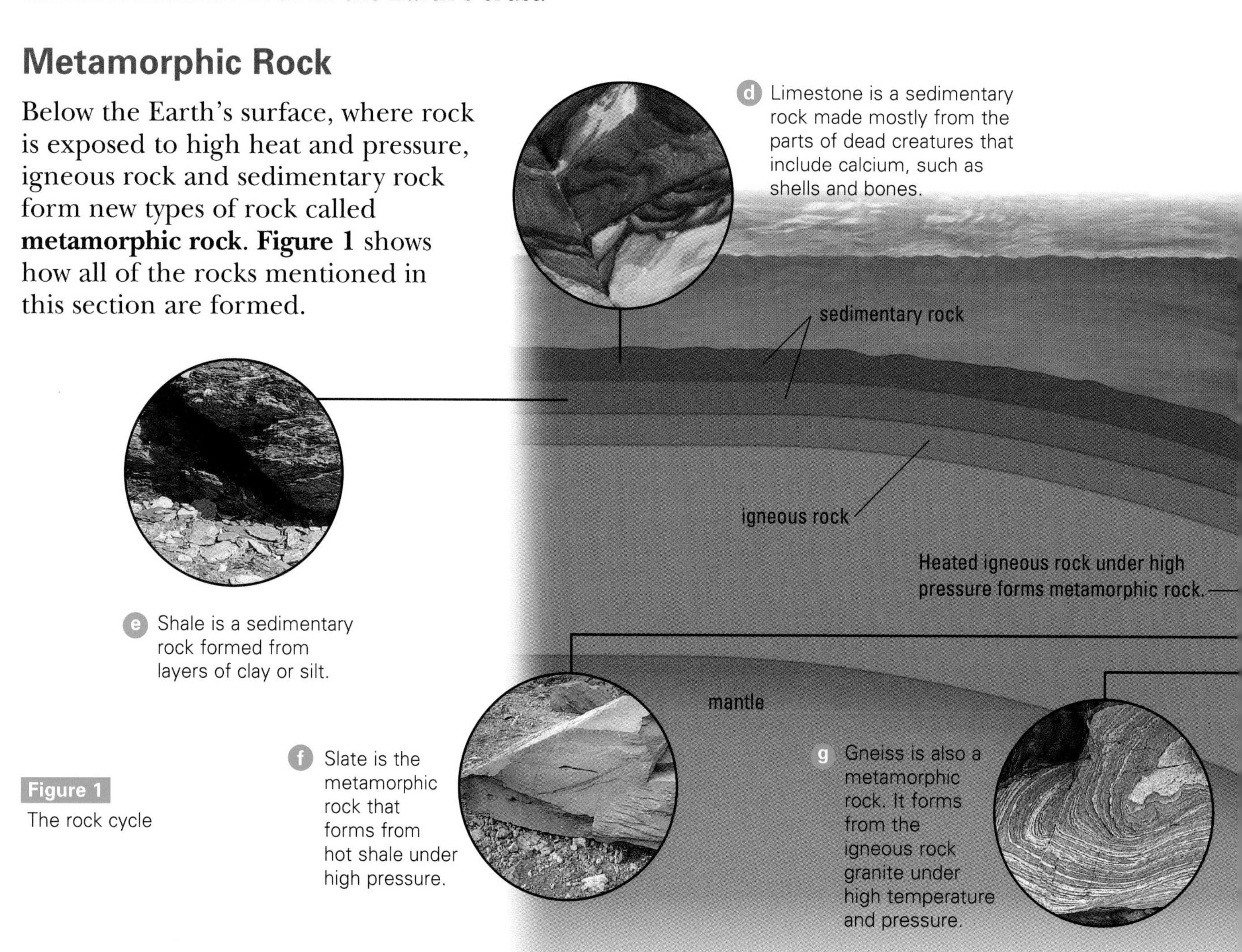

Figure 1 The rock cycle

The Rock Cycle

Rocks, no matter what type, can gradually change into another type. All three forms of rock—sedimentary, igneous, and metamorphic—may eventually become exposed on the Earth's surface, where erosion will wear them down. The resulting sediment then gradually forms into layers that get compressed into sedimentary rock.

Any rock, if pushed far enough into the Earth, will become metamorphic due to the high temperatures and pressure. If the rock is part of a plate plunging far into the mantle underneath another plate, it will become extremely hot and melt, turning into magma. The magma can, in turn, rise and erupt out of a volcano or cool gradually near the surface, forming igneous rock.

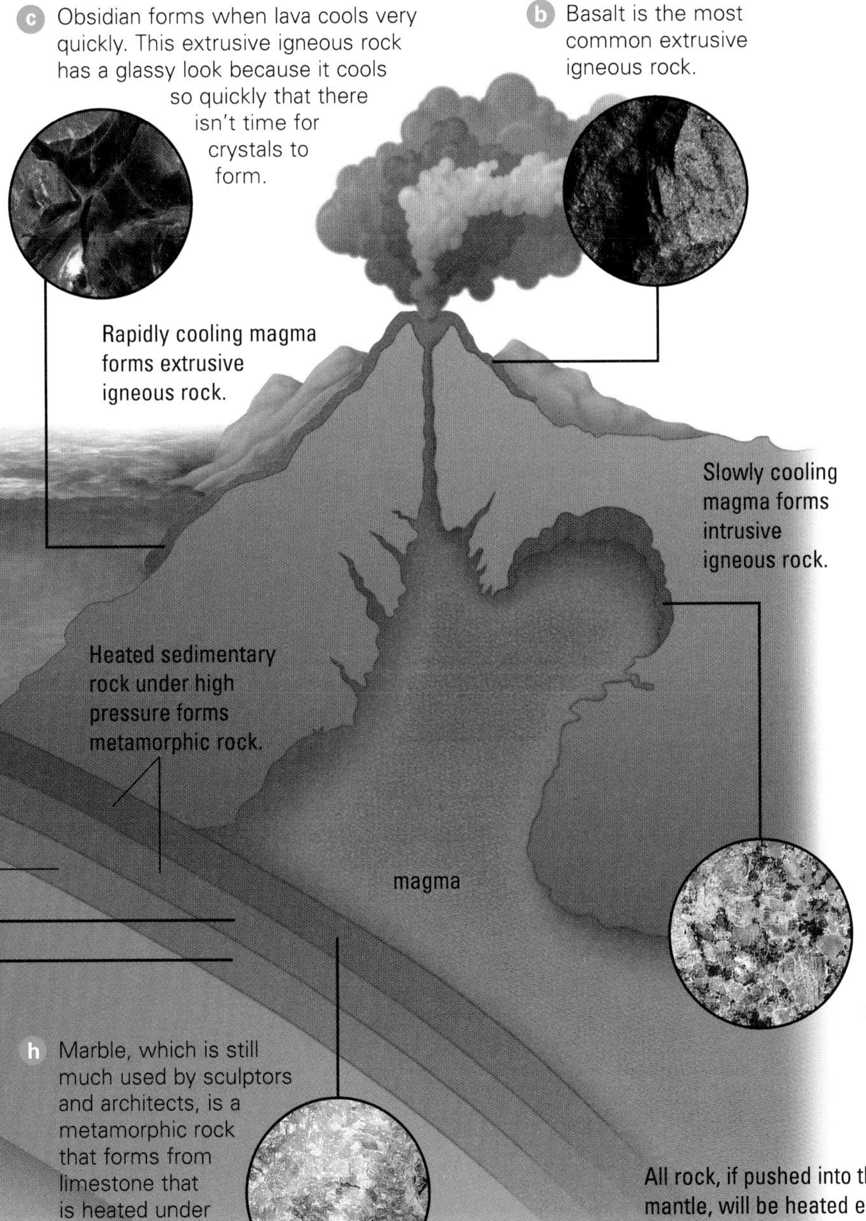

(c) Obsidian forms when lava cools very quickly. This extrusive igneous rock has a glassy look because it cools so quickly that there isn't time for crystals to form.

(b) Basalt is the most common extrusive igneous rock.

(a) Granite is an intrusive igneous rock. Some of the oldest rocks on Earth are granite, such as the Canadian Shield rocks of central and northern Ontario and Quebec.

(h) Marble, which is still much used by sculptors and architects, is a metamorphic rock that forms from limestone that is heated under high pressure.

Understanding Concepts

1. Which type of rock is formed by:
   - **(a)** weathering and erosion?
   - **(b)** pressure and heat, but no melting?
   - **(c)** melting and cooling?
2. What is the difference between intrusive and extrusive igneous rock?
3. Illustrate and describe how particles in a rock could go through the rock cycle from igneous to sedimentary to metamorphic and back to igneous rock.

Making Connections

4. **(a)** Which type of igneous rock was probably the first to form on the early Earth? Explain.
   - **(b)** The most common type of rock on Earth is basalt. Speculate on why this is the case.
5. Coal and diamonds are formed from the same substance—carbon. Diamond is much harder than coal.
   - **(a)** Which one do you think is formed farthest underground? Explain.
   - **(b)** Speculate on how these differences might affect mining.

Exploring

6. Before the invention of modern materials, slate used to have many uses because it is a hard rock that cleaves into sheets. Research what some of these uses were and what materials have replaced slate. What advantages do modern materials have over slate?

Volcanoes: Mountains from Magma

The Earth's surface is pockmarked with volcanoes. Some of them erupt frequently and relatively quietly, and you can actually watch the lava flow out of them from a safe distance. Others erupt only once every few hundred years, but when they do, it is wise to be as far away as possible.

Shield Volcanoes: Rivers of Lava

Shield volcanoes do not occur at the edge of plates, unlike the major mountain ranges of the world. Instead, they can be found anywhere in a plate, even rising out of the ocean floor. **Shield volcanoes** are formed above hot spots in the mantle, as shown in **Figure 1**. In a hot spot, magma collects in enormous pools. The hot magma eventually melts the rock above it and pours out through the hole onto the Earth's surface as **lava**. The lava of shield volcanoes tends to be runny and hardens into basalt rock. Basalt lava is so fluid that shield volcanoes tend not to erupt explosively. The lava simply pours out of the volcano like a river and then hardens. Because of this, shield volcanoes build up gradually and have gently sloping sides.

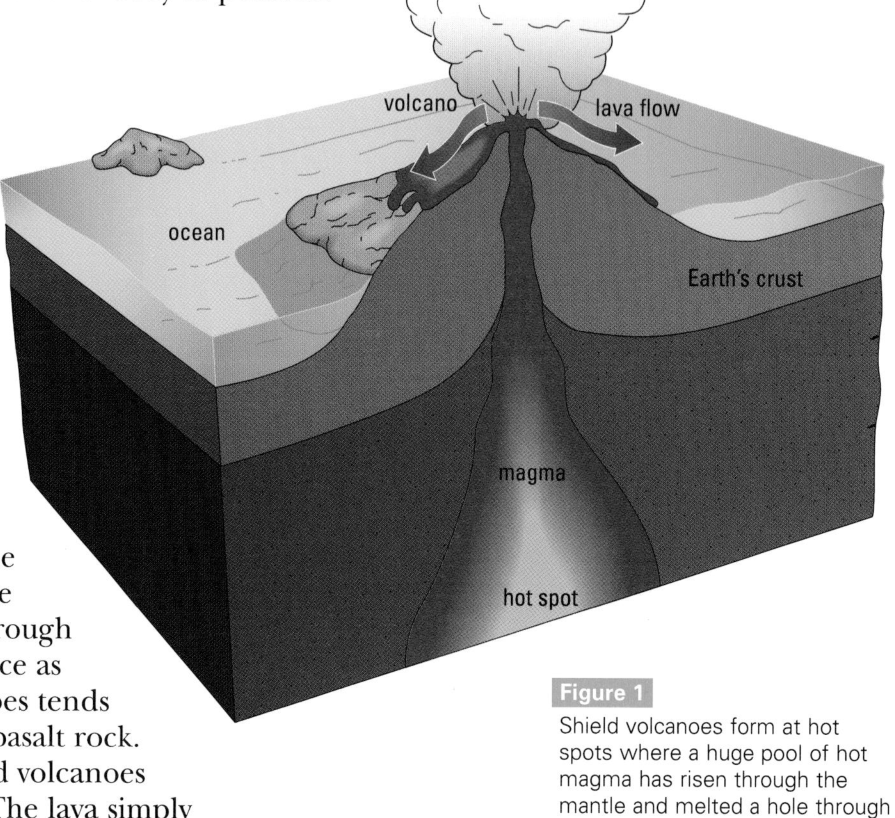

Figure 1
Shield volcanoes form at hot spots where a huge pool of hot magma has risen through the mantle and melted a hole through the solid rock of the Earth's crust.

When a shield volcano forms on the ocean floor, the lava pouring out hardens more quickly than it would on land. More lava pours on top, forming a volcanic **cone**. The cone may build up until it rises above sea level and forms an island. **Figure 2** shows Mauna Loa, a shield volcano that has formed the island of Hawaii.

Figure 2
Mauna Loa, on the island of Hawaii, is a shield volcano. From its base at the bottom of the Pacific Ocean to its summit, it rises 9750 m, which makes it taller than Mount Everest.

SKILLS HANDBOOK: (8A) Writing a Report (4A) Research Skills

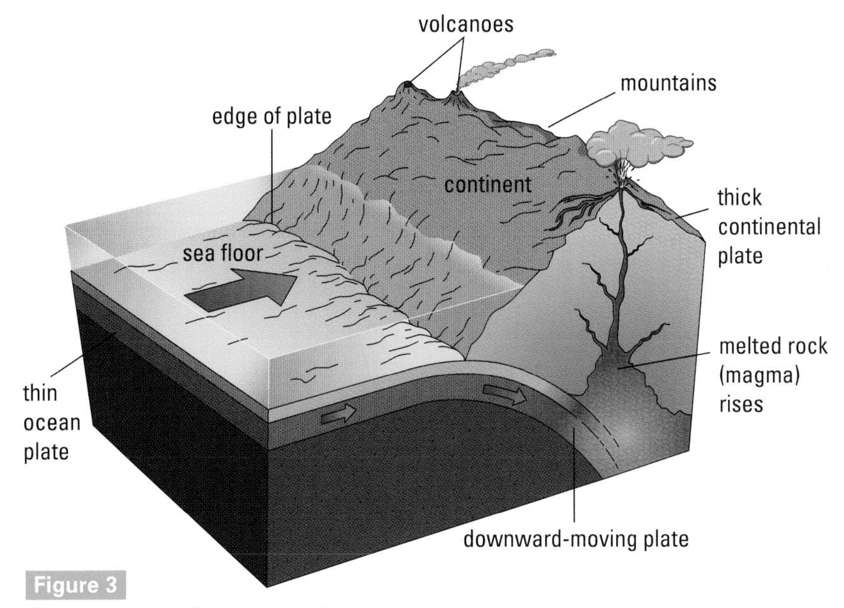

Figure 3

Because the sediment that slides under the continental plate contains water, the lava of stratovolcanoes tends to be explosive.

Stratovolcanoes: Volcanoes That Explode

Stratovolcanoes are the type of volcanoes that we usually hear about in the news because they erupt so explosively, blowing ash and rock many kilometres up into the atmosphere, along with an enormous column of steam and gases.

As shown in **Figure 3**, this type of volcano forms where two plates collide, one sliding on top of the other. As the light sediments on the descending plate heat up and melt, the resulting magma rises up against the plate above and melts through it. This forms a hole so the magma can escape.

However, in contrast to the very liquid magma of shield volcanoes, the magma of stratovolcanoes is thick and sticky. Because the magma is made of melted crust and the sediment on the lower plate contains water, the lava of stratovolcanoes also contains steam under enormous pressure. It is this combination of thick magma and steam that causes stratovolcanoes to erupt so explosively. As the lava rises, the high-pressure steam escapes, carrying lava and ash with it. Mount St. Helens, shown in **Figure 4**, is a stratovolcano.

Understanding Concepts

1. Which type of volcano has very liquid magma?
2. Where are stratovolcanoes found? Why?
3. Where do eruptions occur without forming volcanoes?

Making Connections

4. Take a look at the world map of plates in section 4.16. On the basis of your information about plate movement, do you think that Mount Vesuvius would produce dangerous volcanic eruptions or quiet eruptions?
5. As plates move apart, the oceans between them expand. The Atlantic Ocean is expanding this way. Based on what you know, what rock would you expect makes up most of the ocean floor?

Exploring

6. (8A) Some famous volcanic eruptions had devastating consequences for the people who lived near them. Prepare a report on one of the following eruptions or on one of your own choice: Thera in about 2600 BC; Mount Pelee in AD 1902.
7. (4A) The largest volcanic eruptions in the past have given off so much dust and ash that they have formed a dust veil in the atmosphere that covered the entire planet. Research whether these dust veils made Earth's climate warmer or cooler and why.

Figure 4

Mount St. Helens is a stratovolcano in the Cascade Mountains of Washington State. In 1980, it erupted for the first time in hundreds of years.

Eruptions Where Plates Part

As you have already seen, lava also flows out onto the Earth's surface where plates are moving apart, usually on the ocean floor where the crust is thinnest. Here the lava cools quickly to form basalt in a long ridge on each side of the crack in the ocean floor. Some of these ridges may rise high enough to reach the surface, creating islands such as Iceland.

- Recording
- Evaluating
- Communicating

Design and Build a Model for Using the Earth Responsibly

Human activities affect the Earth's crust. When we dig for metals and other minerals, when we dump waste on the land, or when we clear land and use it to grow crops, we make changes in the crust. Sometimes those changes cannot be reversed. When we use land, we should use it responsibly, keeping the impact on other living things to a minimum. In completing these challenges, you will explore responsible land use.

Figure 3

Farming involves intensive use of the land. It must be carefully planned.

Figure 1

Open-pit mines require extensive landscaping when they are no longer in use.

1 A Responsible Mine

Problem situation

Geologists have identified a body of rock with copper-containing ore 600 m below the surface in northwestern Ontario. The ore is under an old-growth pine forest near a provincial park.

Design brief

- Design and build a model of a mine that both removes ore efficiently and causes minimal environmental damage.

Design criteria

- In the model, rock must be safely removed from the rock face.
- The model must include a mechanism that brings rock containing copper ore from the rock face to the surface.
- The mine must be environmentally friendly. Damage to the forest and the nearby park must be kept to a minimum.
- Your model must include a plan for restoring the land once the mine is closed.
- Your model mine must be safe for miners.

2 A Safe Mine-Tailings Pond

Problem situation

After rock is brought to the surface of a mine, the desired mineral, such as chalcopyrite (which contains copper), must be separated from the rock. Usually this involves crushing the rock and mixing it with water and chemicals. Then the valuable mineral is skimmed off. The remaining wastes, called tailings, are poured into a large artificial pond. Tailings are often toxic and must not be allowed to escape into the surrounding soil and water systems.

Design brief

- Design and build a model of a tailings pond that will safely contain all the tailings from a mine until they can be treated and removed.

Design criteria

- The model tailings pond must be able to contain both toxic solids and liquids.
- Plans for testing surrounding land and water for contamination must be included.
- The model must demonstrate how the contents of the pond can be removed without affecting the environment.
- The model must include a plan for restoring the land used for the pond to its original state.

3 An Erosion-Proof Field

Problem situation

The world's human population of approximately 6 billion relies on farmers to grow food. Poor farming techniques can result in precious topsoil being washed away by rain or blown away in windstorms. Chemicals used by farmers, such as pesticides, herbicides, and fertilizers, can pollute nearby streams and rivers.

Design brief

- Design and build a model of a farmer's field in which the soil resists erosion and water runoff.

Design criteria

- The model field should have a slope of 15° with a stream at its lowest point.
- The model must show two sections: one at the preplanting stage (no plants yet) and one that is already planted.
- When the model field is sprinkled with water from a watering can, no soil should reach the stream at the bottom. You must create a test to show that soil does not enter the stream after the sprinkling.

Assessment

Your model will be assessed according to how well you:

Process

- understand the problem
- develop a safe plan
- choose and safely use appropriate materials, tools, and equipment
- test and record results
- evaluate your model, including suggestions for improvement

Communication

- prepare a presentation
- use correct terms
- write clear descriptions of the steps you took in building and testing your model
- explain clearly how your model solves the problem
- make an accurate technical drawing for your model

Product

- meet the design criteria with your model
- use your chosen materials effectively
- construct your model
- solve the identified problem

When preparing to build or test a design, have your plan approved by your teacher before you begin.

Figure 2

Tailings ponds are temporary dumping sites for waste from mining.

Unit 4 Summary

In this unit you have learned that the Earth's crust—the rocks and soil that sustain us—is constantly changing as many forces act on it, including those set in motion by human beings.

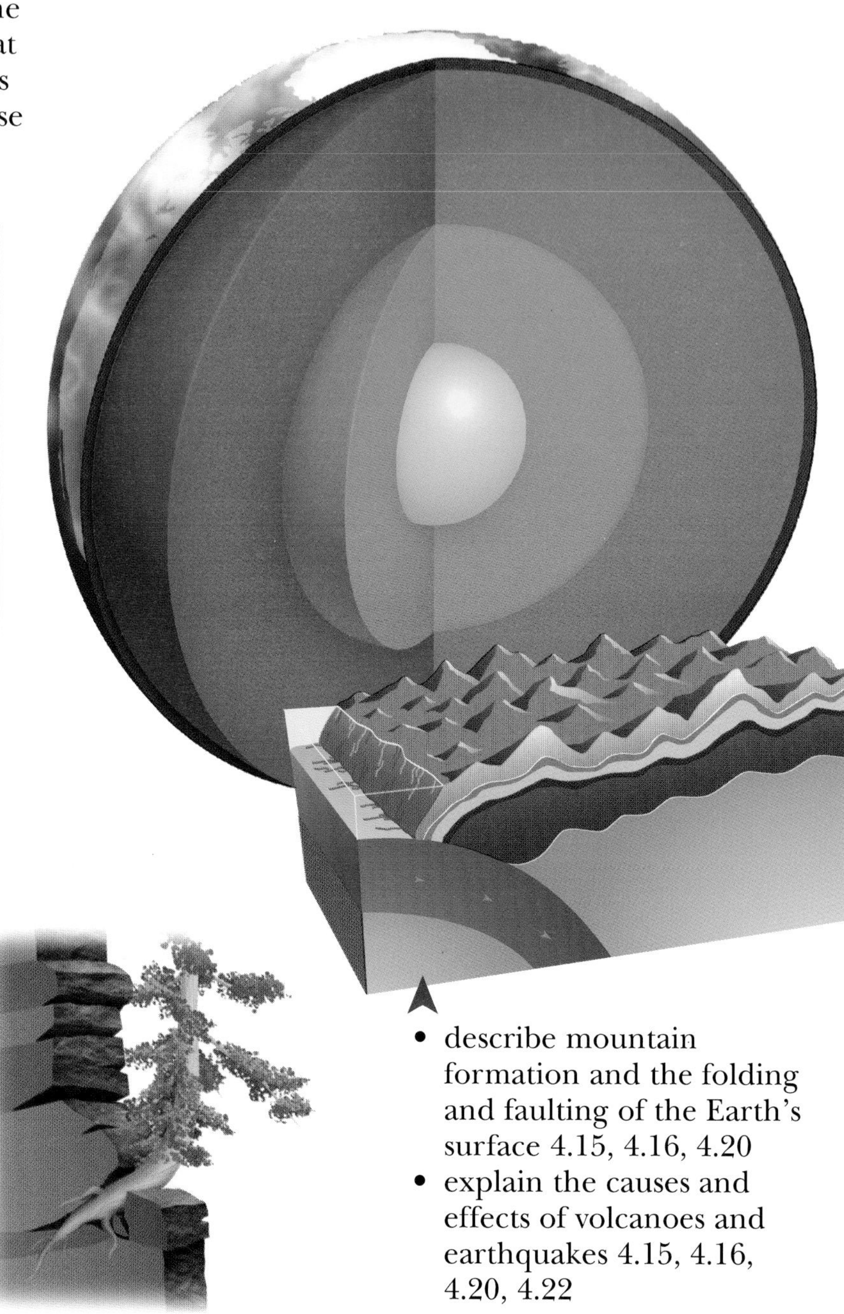

Reflecting

- Reflect on the ideas and questions presented in the Unit Overview and in the Getting Started. How can you connect what you have done and learned in this unit with those ideas and questions? (To review, check the sections indicated in this Summary.)
- Revise your answers to the Reflecting questions in 1, 2, 3 and the questions you created in the Getting Started. How has your thinking changed?
- What new questions do you have? How will you answer them?

Understanding Concepts

- describe the Earth and its crust 4.1, 4.15
- distinguish between rocks and minerals 4.2
- classify rocks and minerals by their characteristics and by how they were formed 4.2, 4.13, 4.21
- describe how soil is formed 4.6, 4.7
- describe the origin and history of natural features of the local landscape 4.6, 4.11, 4.13, 4.20
- observe and analyze evidence of geological change 4.12, 4.14, 4.15
- identify the processes involved in rock and mineral formation 4.13, 4.21
- explain the rock cycle 4.13, 4.21
- describe mountain formation and the folding and faulting of the Earth's surface 4.15, 4.16, 4.20
- explain the causes and effects of volcanoes and earthquakes 4.15, 4.16, 4.20, 4.22

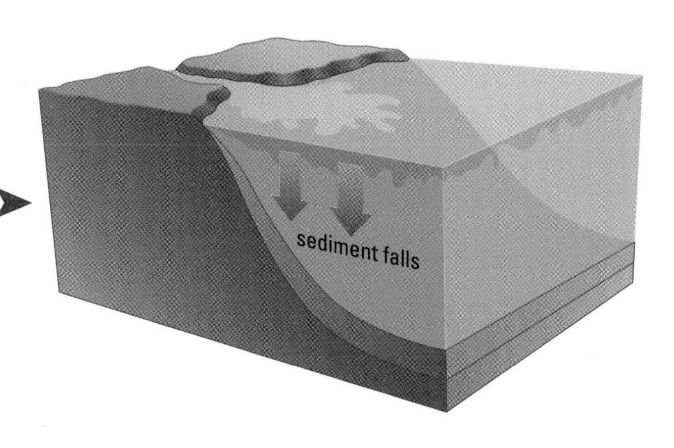

Applying Skills

- investigate the effect of weathering on rocks and minerals 4.6, 4.11, 4.12, 4.13
- classify minerals, using observations, according to their characteristics 4.2
- describe the process of mineral extraction from rock 4.3, 4.4
- observe what makes topsoils different from one another 4.8
- understand and use the following terms:
- design and build a mountain that withstands the forces of erosion 4.12
- plot historical data of earthquake and volcanic activity in order to determine patterns and predict future events 4.17
- design, plan, and carry out the construction of an earthquake-proof building 4.19

bedrock	minerals
clay	mining (strip and underground)
cleavage	mountains (block, dome, and fold)
colour	moraine
continental drift	ore
cone	overburden
core (inner and outer)	plates
crop rotation	plate tectonics
crust	ridge
deposit	sand
drift	sediment
drumlin	sedimentary rock
erosion	shield volcanoes
fault	silt
flood plain	stratovolcanoes
hardness	subduction
horizons	subsoil
humus	tailings
igneous rock (intrusive and extrusive)	till
lava	topsoil
litter	weathering (biological, chemical, and mechanical)
lustre	zero tillage
magma	
mantle	
metamorphic rock	

Making Connections

- investigate ways in which humans have altered the landscape to meet their needs 4.3
- identify factors that must be considered when making informed decisions about land use 4.4, 4.5, 4.10
- investigate soil to determine its suitability for specific uses, including conservation 4.7, 4.8, 4.9, 4.10

- identify a career that uses modern technology to contribute to the study of natural geological events 4.18, 4.19

Unit 4 Review

Understanding Concepts

1. (a) Describe the structure of the Earth.
   (b) Even though the inner layers of the Earth are extremely hot, the inner core is solid. What is the reason?
2. What is the difference between a rock and a mineral?
3. Which of the following metals are not minerals: copper, bronze, gold, tin, steel, silver? Give a reason for your answer.
4. (a) What property of minerals is assessed using a scratch test?
   (b) Describe how you could use a penny in a scratch test of three minerals.
5. Why do all minerals cleave along flat surfaces when they are broken up?
6. If two samples of the same mineral have crystals of different sizes, what does this tell you about the samples?
7. (a) Explain how lichen erodes rock.
   (b) What other types of biological weathering can erode rock?
8. Ice causes erosion in two ways. What are they?
9. Is each of the following an example of mechanical or chemical weathering or both? Explain your answer.
   (a) cracks in the sidewalk
   (b) discoloured metal jewellery found in a sunken ship in the ocean
   (c) an underground cave
10. Which area would be more likely to lose its topsoil: a gently sloping area or a steep hill? Explain.
11. List the following soil particles from largest to smallest: sand, clay, silt, pebbles.
12. Why do gardeners mix sand and peat into soils that are mostly clay?
13. Surtsey, an island near Iceland, was formed recently from lava flowing out of the ridge in the middle of the Atlantic Ocean. Explain, using diagrams, how deep soil might eventually form on this island.
14. Explain how you can classify soils using your senses of touch and sight.
15. Look at **Figure 1**.
    (a) Which river is younger, a or b?
    (b) Which river is most likely to carry large stones?
    (c) Which river is most likely to be surrounded by a flood plain?

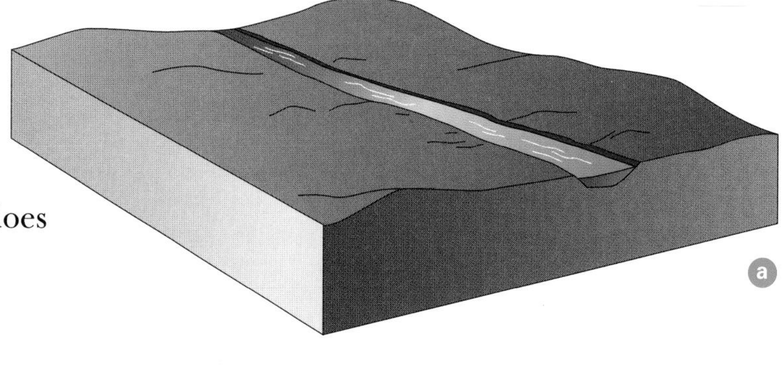

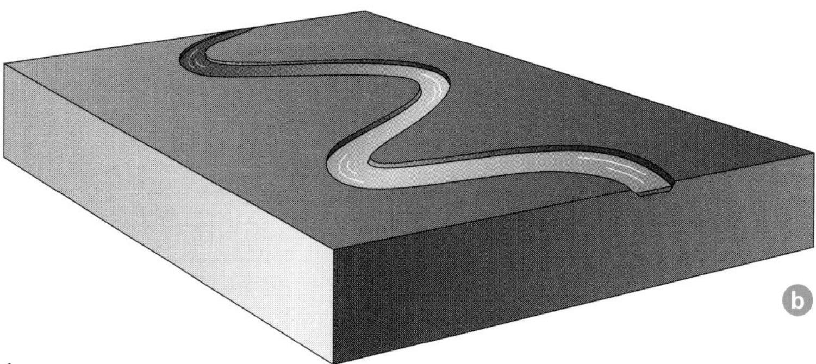

Figure 1

16. A small lake near a river has a "C" shape and a sandy bottom.
    (a) Could this lake have once been part of the river?
    (b) What other clues would you look for to support your answer?
17. In what type of rock are fossils normally found? Why?

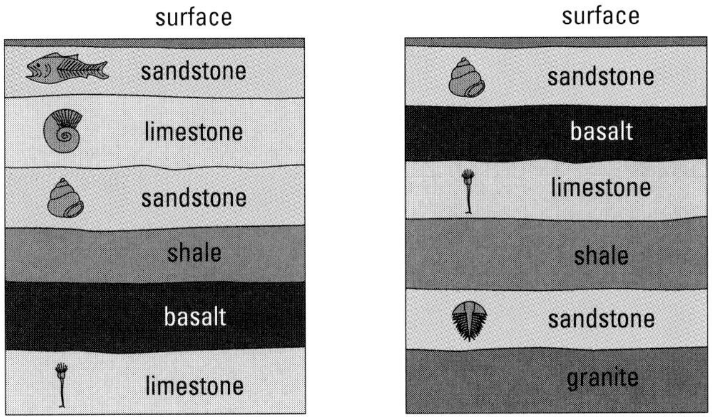

surface

(bony fish)	sandstone
(ammonite)	limestone
(snail)	sandstone
	shale
	basalt
(crinoid)	limestone

surface

(snail)	sandstone
	basalt
(crinoid)	limestone
	shale
(trilobite)	sandstone
	granite

	Fossil	Epoch
	bony fish	Cenozoic
	ammonite	Jurassic
	snail	Triassic
	crinoid	Permian
	trilobite	Cambrian

Figure 2

18. **Figure 2** shows layers of rock in two locations. Fossils have been found in some of the layers.

(a) Is it possible that the layers of basalt in the two locations were produced by the same volcano? Explain.

(b) Which is older, the layer of shale in Location 1 or the layer of shale in Location 2? Explain your deduction.

(c) Explain the differences in the rocks from the two locations.

19. Explain how tectonic plate movements cause:

(a) earthquakes.

(b) volcanoes.

20. The Laurentian Mountains in Quebec are low and rounded. They were not always this shape.

(a) Describe how this mountain range may have appeared 200 million years ago. What has happened to them since then?

(b) Based on your understanding of mountains, speculate on how this mountain range formed.

21. What is the difference between intrusive and extrusive igneous rock?

22. Lava flows down the side of a shield volcano and enters ocean water.

(a) What kind of rock will be formed from the lava?

(b) Would you expect the mineral crystals in the rock to be large or small?

23. India and the island of Sri Lanka are on the Indo-Australian Plate, which, in the north, is being subducted under the much larger Eurasian Plate.

(a) Would you expect the collision of these two plates to cause earthquakes in Sri Lanka? Explain.

(b) Draw a diagram showing what is happening where the two plates meet.

(c) On the Indo-Australian plate there are areas where layers of sandstone lie on top of granite. Explain what will happen to this rock as it is subducted.

24. Imagine that you could mark a tiny crystal of mineral in a large rock sitting on the surface in central Ontario. Imagine also that you could go away for a billion years, come back, and find that crystal. List as many places as you can think of where that crystal might end up. For each place, explain how the crystal could get there.

Applying Skills

25. Match the following terms with the correct descriptions:

Description	Term
A place where two plates slide past each other	1 mantle
B molten rock on the Earth's surface	2 moraine
C partly molten layer below the Earth's crust	3 fault
D molten rock below the Earth's surface	4 lava
E deposit of gravel and loose rock left by glaciers	5 sediment
F rock particles deposited by moving water	6 magma

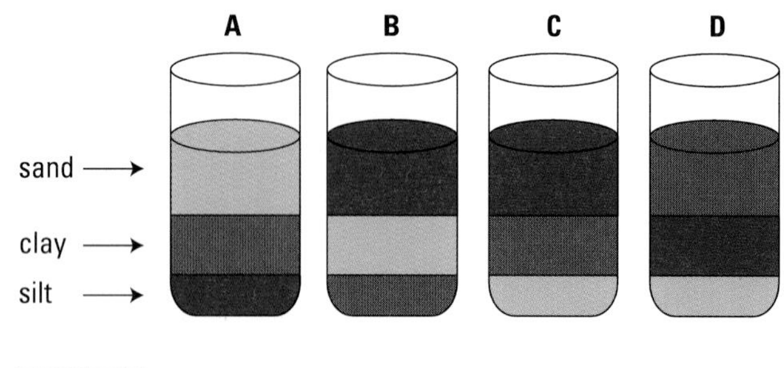

Figure 3

26. The layers produced by four soil samples are shown in **Figure 3**.
    (a) Which illustration shows an accurate representation of the layering you would expect?
    (b) Which soil sample would you recommend for planting vegetables? Explain.

27. You want to start a garden in an area where there has never been one before.
    (a) Only a few weeds seem to be growing in the soil. What does this indicate?
    (b) You notice that puddles form on the surface of the soil. What does this indicate?
    (c) What could you do to improve the soil?

28. **Figure 4** shows rocks from three different areas. In each case, identify whether the rocks are igneous, metamorphic, or sedimentary, and explain your reasoning.

29. A geologist notices that small, narrow hills in one area seem to point in the same direction. When she takes samples from them, she finds that there is gravel near the surface of each one.
    (a) What type of erosion occurred here?
    (b) Did the erosion occur once over a long period of time, or at several different times? Explain.

30. "Fossils give an incomplete history of life on Earth." Is this statement true or false? Explain.

31. Why do earthquakes often occur in the same geographical regions as volcanoes?

32. A travel agent finds customers are fascinated by volcanoes. He wants to arrange a trip that will visit an active volcano. He has read predictions that volcanoes may erupt soon in Mexico, Alaska, and Iceland.
    (a) Which volcano would you advise him is safest to visit? Why?
    (b) The travel agent thinks the volcano in Alaska is safe because volcanoes in that state have caused little loss of life in the past. Is the agent's reasoning correct? Why or why not?

33. A farmer is spending a lot of money on herbicides and insecticides and is wondering what she can do to spend less. Suggest some ways she could reduce her costs. Illustrate and explain your suggestions.

34. A small earthquake occurs in southern Ontario and people are worried that it's a sign of a much bigger one to come. Is it reasonable to worry? Explain.

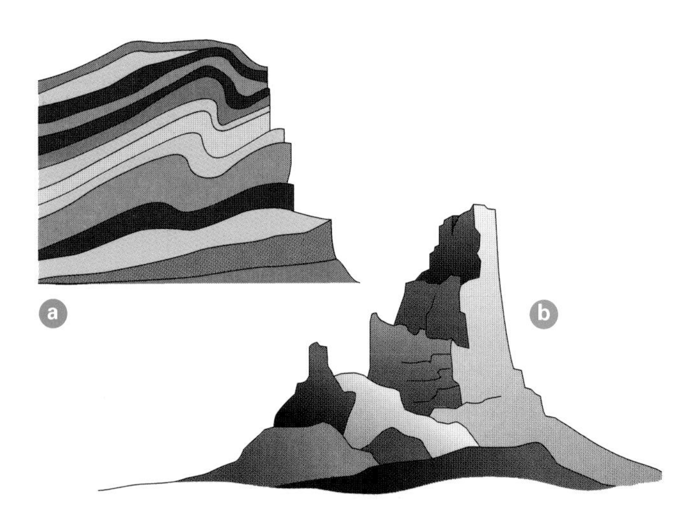

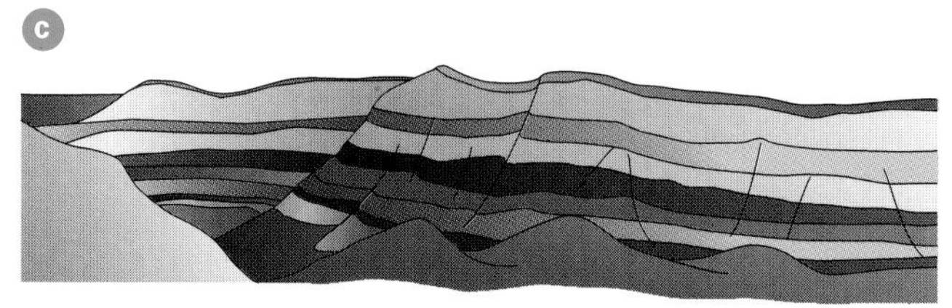

Figure 4

Making Connections

35. What factors help determine the value of a mineral?

36. If a huge diamond mine were discovered, so diamonds become common rather than rare, how might the use of diamonds change?

37. A person from Norway visits Newfoundland and notices the local hills look just like those at home. Explain the resemblance.

38. Imagine that you are planning to build a home on a steep hillside where several lots are available.

(a) What should you be concerned about when choosing a lot?

(b) What steps should you take to prevent erosion if you decide to build on a sloped lot?

(c) How might landscaping with plants help prevent erosion?

39. The land on a flood plain makes particularly good farmland.

(a) Why do you think this is the case?

(b) Is a flood plain a safe place for people to live? Explain.

40. You decide to go into the mineral mining business. After much study, you discover local deposits of five minerals. You have enough resources to mine only one. In **Table 1** these five minerals are ranked for abundance (the higher the number, the larger the deposit); ease of mining (the higher the number, the easier it will be to mine); and the value of the mineral (the higher the number, the more valuable the mineral). Which mineral would you mine? Why?

41. A woman is planning to buy a farm. When she goes to look at farms that are for sale she brings a hand lens and a shovel with her. How would she use these tools to help her decide which farm to buy?

42. Look at **Figure 5**.

(a) Using plate tectonics, explain the feature shown.

(b) East of this area there is a continent. What features would you expect to find near the coast?

(c) Many people are moving into cities along the coast. What advice would you give them? Why?

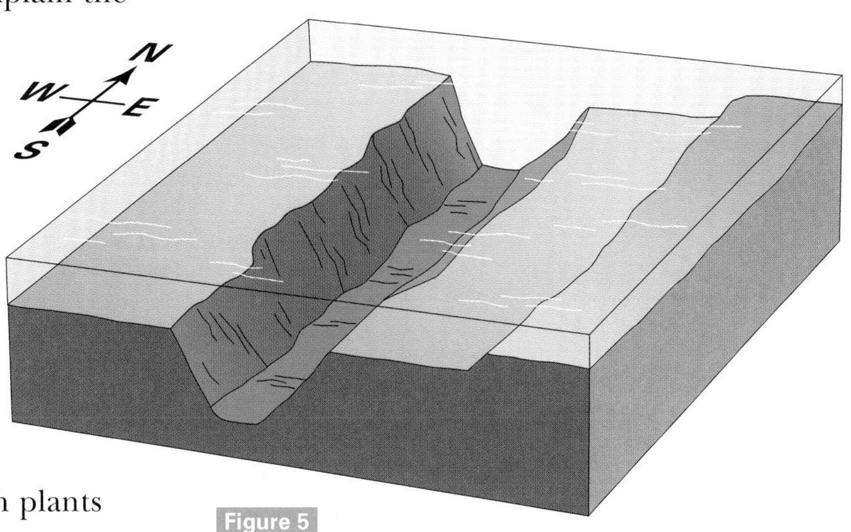

Figure 5

43. An ancient statue from a small village in Egypt, almost perfectly preserved, was moved to a large city in the United States, where it quickly eroded. What different conditions in the city might have caused this to occur? How would you reduce the erosion?

44. Using terms you've learned in this unit, design a concept map that illustrates the processes that change the Earth's crust. There is a list of terms in the Unit Summary.

Table 1

Mineral	Abundance	Ease of mining	Value
Bauxite	5	4	3
Hematite	4	5	1
Halite	3	3	2
Quartz	2	1	4
Gold	1	2	5

Glossary

B

bedrock: the layer of rock immediately below the subsoil

C

clay: soil component that consists of very small particles of rock (less than 0.002 mm)

cleavage: a property of minerals; describes the way the minerals split into smaller pieces

colour: a property of minerals; describes the appearance of the mineral; used to identify and classify minerals

cone: the shape that hardened lava takes as it spills from a volcano

continental drift: a hypothesis that continents move and that millions of years ago they were all connected into a supercontinent

core, inner: the innermost region of the Earth and also its hottest; made mostly of solid iron and nickel

core, outer: the region of the Earth surrounding the inner core made of liquid iron and nickel

crop rotation: a process used by farmers to prevent loss of nutrients from soil in which in each growing season, a different kind of plant is grown in a field

crust: the thin surface layer of rock that covers the Earth

D

deposit: an area in which the local rock contains unusually large amounts of a valuable mineral or metal

drift: loose material that is easily scraped from the bedrock by an advancing glacier; eventually left behind when the glacier recedes

drumlins: small mounds of moraine shaped by glaciers

E

erosion: the wearing away of soil or rock by agents such as wind, water, and living things

F

fault: an area where rocks are broken by movement in the crust

flood plain: the area on each side of a river that is covered when the river overflows its banks, mostly during the spring runoff

H

hardness: a property of minerals; describes how resistant a mineral is to being scratched

horizons: the layers that can be found in most soil, from litter to bedrock

humus: decaying plant and animal matter that is mixed with soil

I

igneous rock: rock formed from the hardening of liquid magma

igneous rock, intrusive: rock formed when liquid magma cools and hardens below the Earth's surface

igneous rock, extrusive: rock formed when liquid magma cools and hardens above the surface of the Earth

L

lava: magma that flows out of cracks onto the Earth's crust

litter: material lying on the surface of soil, consisting of leaves, broken branches, fallen trees, and animals' bodies

lustre: a property of minerals; describes how shiny a mineral is

M

magma: a hot, liquid solution of dissolved minerals; cools to form igneous rock

mantle: the area between the Earth's crust and its core; it consists of magma, a thick, molten material that, when cooled, can form rock

metamorphic rock: rock formed when igneous or sedimentary rock is exposed to the higher temperatures and pressure deep in the Earth's crust

minerals: the building blocks of rocks; minerals are non-living things

mining, strip: a type of mining whereby topsoil and rock are removed from the top of a deposit so that a mineral can be removed; used only for deposits that are near the surface

mining, underground: a type of mining whereby one or many tunnels are dug into rock to reach a deposit so that a mineral can be removed; often used for deposits that are buried far below the surface

moraines: piles of broken rock left behind by receding glaciers

mountains, block: mountains formed when movements in the crust cause blocks of crust to rise or fall

mountains, dome: mountains formed by the pressure of hot magma pushing up from below

mountains, fold: mountains formed when two of the Earth's plates collide

O

ore: rock that contains a valuable mineral

overburden: top layer of soil and rock which must be removed during strip mining to expose ore

P

plates: solid sections of the Earth's crust that float on the liquid mantle

plate tectonics: a theory that describes the Earth's crust as a set of rocky plates that are in continual motion, colliding and moving apart

R

ridge: a rise made of hardened lava erupting where two of the Earth's plates are moving apart; usually found on the ocean floor

S

sand: soil component that consists of largest particles of rock (0.02–2.0 mm)

sediment: fine particles of rock and soil that are deposited by moving water; settles in layers on lake beds or sea beds

sedimentary rock: rock composed of layers of sediment

shield volcanoes: volcanoes formed above a hot spot in the mantle of the Earth; eruptions tend to be frequent but not spectacular

silt: soil component that consists of medium-sized particles of rock (0.002–0.02 mm)

stratovolcanoes: volcanoes formed on the upper surface of a plate where another plate is plunging underneath; eruptions tend to be dramatic and destructive

subduction: a process whereby one plate slides beneath another plate, pushes into the hot mantle, heats up, and melts

subsoil: the layer of soil below the topsoil; contains very little humus and many stones

T

tailings: liquid waste from a mine, consisting of a mixture of small particles of rock and the chemicals used to extract the valuable mineral from ore

tailings pond: a pond built to contain toxic liquid wastes from mining

till: to break up compacted soil with a plough in order to allow water and air into the soil; prepares the soil for plant growth

topsoil: the layer of soil under the litter; contains humus and small amounts of stones

W

weathering, biological: erosion caused by living things

weathering, chemical: erosion caused by chemicals, such as the acid in acid rain

weathering, mechanical: erosion caused by fast-moving water, particles carried by the wind, or ice

Z

zero tillage: a method of farming whereby stubble from previous crop is left in the ground, and new seed is planted into old stubble, without the use of a plough; reduces loss of soil nutrients

Index